Textbook Outlines, Highlights, and Practice Quizzes

Statistics: The Art and Science of Learning from Data

by Alan Agresti, 3rd Edition

All "Just the Facts101" Material Written or Prepared by Cram101 Publishing

Title Page

STUDYING MADE EASY

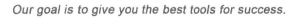

This Cram101 notebook is designed to make studying easier and increase your comprehension of the textbook material. Instead of starting with a blank notebook and trying to write down everything discussed in class lectures, you can use this Cram101 textbook notebook and annotate your notes along with the lecture.

Our goal is to give you the best tools for success.

For a supreme understanding of the course, pair your notebook with our online tools. Should you decide you prefer Cram101.com as your study tool,

we'd like to offer you a trade...

Our Trade In program is a simple way for us to keep our promise and provide you the best studying tools, regardless of where you purchased your Cram101 textbook notebook. As long as your notebook is in *Like New Condition**, you can send it back to us and we will immediately give you a Cram101.com account free for 120 days!

Let The **Trade In** Begin!

THREE SIMPLE STEPS TO TRADE:

1. Go to www.cram101.com/tradein and fill out the packing slip information.

2. Submit and print the packing slip and mail it in with your Cram101 textbook notebook.

3. Activate your account after you receive your email confirmation.

* Books must be returned in *Like New Condition*, meaning there is no damage to the book including, but not limited to; ripped or torn pages, markings or writing on pages, or folded / creased pages. Upon receiving the book, Cram101 will inspect it and reserves the right to terminate your free Cram101.com account and return your textbook notebook at the owners expense.

Visit Cram101.com for full Practice Exams

"Just the Facts101" is a Cram101 publication and tool designed to give you all the facts from your textbooks. Visit Cram101.com for the full practice test for each of your chapters for virtually any of your textbooks.

Cram101 has built custom study tools specific to your textbook. We provide all of the factual testable information and unlike traditional study guides, we will never send you back to your textbook for more information.

YOU WILL NEVER HAVE TO HIGHLIGHT A BOOK AGAIN!

Cram101 StudyGuides

All of the information in this StudyGuide is written specifically for your textbook. We include the key terms, places, people, and concepts... the information you can expect on your next exam!

Want to take a practice test?

Throughout each chapter of this StudyGuide you will find links to cram101.com where you can select specific chapters to take a complete test on, or you can subscribe and get practice tests for up to 12 of your textbooks, along with other exclusive cram101.com tools like problem solving labs and reference libraries.

Cram101.com

Only cram101.com gives you the outlines, highlights, and PRACTICE TESTS specific to your textbook. Cram101.com is an online application where you'll discover study tools designed to make the most of your limited study time.

By purchasing this book, you get 50% off the normal subscription free!. Just enter the promotional code **'DK73DW20551'** on the Cram101.com registration screen.

www.Cram101.com

Learning System

Statistics: The Art and Science of Learning from Data
Alan Agresti, 3rd

CONTENTS

facts101

CHAPTER OUTLINE: KEY TERMS, PEOPLE, PLACES, CONCEPTS

_____ | Categorical variable

_____ | Contingency table

_____ | Chi-squared test

_____ | Statistic

_____ | T-test

_____ | Probability

_____ | Statistical inference

_____ | Changing-criterion research design

_____ | Inference

_____ | Internet

_____ | Sample survey

_____ | Squared

_____ | Subject

_____ | Correlation

_____ | Bar graph

_____ | Descriptive statistic

_____ | Eurobarometer

_____ | Opinion

_____ | Margin of error

	Parameter
	ANOVA
	Random sampling
	Randomness
	Database
	Variance

CHAPTER HIGHLIGHTS & NOTES: KEY TERMS, PEOPLE, PLACES, CONCEPTS

Categorical variable	Categorical Variables
	In statistics, a categorical variable is a variable that can take on one of a limited, and usually fixed, number of possible values. Categorical variables are often used to represent categorical data.
	A categorical variable that can take on exactly two values is termed a binary variable and is typically treated on its own as a special case.
Contingency table	In statistics, a contingency table is a type of table in a matrix format that displays the (multivariate) frequency distribution of the variables. It is often used to record and analyze the relation between two or more categorical variables.
	The term contingency table was first used by Karl Pearson in 'On the Theory of Contingency and Its Relation to Association and Normal Correlation', part of the Drapers' Company Research Memoirs Biometric Series I published in 1904.
Chi-squared test	A chi-squared test, also referred to as chi-square test or

χ^2 test, is any statistical hypothesis test in which the sampling distribution of the test statistic is a chi-squared distribution when the null hypothesis is true, or any in which this is asymptotically true, meaning that the sampling distribution (if the null hypothesis is true) can be made to approximate a chi-squared distribution as closely as desired by making the sample size large enough.

Some examples of chi-squared tests where the chi-squared distribution is only approximately valid:•Pearson's chi-squared test, also known as the chi-squared goodness-of-fit test or chi-squared test for independence. When mentioned without any modifiers or without other precluding context, this test is usually understood .•Yates's correction for continuity, also known as Yates' chi-squared test.•Cochran-Mantel-Haenszel chi-squared test.•McNemar's test, used in certain 2 × 2 tables with pairing•Linear-by-linear association chi-squared test•The portmanteau test in time-series analysis, testing for the presence of autocorrelation•Likelihood-ratio tests in general statistical modelling, for testing whether there is evidence of the need to move from a simple model to a more complicated one (where the simple model is nested within the complicated one).

One case where the distribution of the test statistic is an exact chi-squared distribution is the test that the variance of a normally distributed population has a given value based on a sample variance.

| Statistic | A statistic is a single measure of some attribute of a sample (e.g. its arithmetic mean value). It is calculated by applying a function (statistical algorithm) to the values of the items comprising the sample which are known together as a set of data. |

More formally, statistical theory defines a statistic as a function of a sample where the function itself is independent of the sample's distribution; that is, the function can be stated before realisation of the data.

| T-test | A t-test is any statistical hypothesis test in which the test statistic follows a Student's t distribution if the null hypothesis is true. It is most commonly applied when the test statistic would follow a normal distribution if the value of a scaling term in the test statistic were known. When the scaling term is unknown and is replaced by an estimate based on the data, the test statistic (under certain conditions) follows a Student's t distribution. |

| Probability | Probability is a way of expressing knowledge or belief that an event will occur or has occurred. The concept has been given an exact mathematical meaning in probability theory, which is used extensively in such areas of study as mathematics, statistics, finance, gambling, science, Artificial intelligence/Machine learning and philosophy to draw conclusions about the likelihood of potential events and the underlying mechanics of complex systems. |

	Interpretations
	The word probability does not have a consistent direct definition.
Statistical inference	In statistics, statistical inference is the process of drawing conclusions from data subject to random variation, for example, observational errors or sampling variation. More substantially, the terms statistical inference, statistical induction and inferential statistics are used to describe systems of procedures that can be used to draw conclusions from datasets arising from systems affected by random variation, such as observational errors, random sampling, or random experimentation. Initial requirements of such a system of procedures for inference and induction are that the system should produce reasonable answers when applied to well-defined situations and that it should be general enough to be applied across a range of situations.
Changing-criterion research design	In a changing-criterion research design a criterion for reinforcement is changed across the experiment to demonstrate the functional relationship between the reinforcement and the behavior. See Mark Dixon's work with a participant using a short video clip to generate a preference for a progressively delayed variable reinforcement over a fixed shorter delay reinforcement in physical therapy.
Inference	Inference is the act of drawing a conclusion by deductive reasoning from given facts. The conclusion drawn is also called an inference. The laws of valid inference are studied in the field of logic.
Internet	The Internet is a global system of interconnected computer networks that use the standard Internet Protocol Suite (TCP/IP) to serve billions of users worldwide. It is a network of networks that consists of millions of private, public, academic, business, and government networks, of local to global scope, that are linked by a broad array of electronic and optical networking technologies. The Internet carries a vast range of information resources and services, such as the inter-linked hypertext documents of the World Wide Web (WWW) and the infrastructure to support electronic mail.
Sample survey	Sample survey is a survey of a population made by using only a portion of the population.
Squared	In algebra, the square of a number is that number multiplied by itself. To square a quantity is to multiply it by itself. Its notation is a superscripted '2'; a number x squared is written as x^2.
Subject	In library and information science documents (such as books, articles and pictures) are classified and searched by subject - as well as by other attributes such as author, genre and document type. This makes 'subject' a fundamental term in this field.

Chapter 1. Statistics: The Art and Science of Learning from Data

Correlation	In statistics, correlation (often measured as a correlation coefficient, ρ) indicates the strength and direction of a relationship between two random variables. The commonest use refers to a linear relationship. In general statistical usage, correlation or co-relation refers to the departure of two random variables from independence.
Bar graph	A bar chart or Bar graph is a chart with rectangular bars with lengths proportional to the values that they represent. Bar charts are used for comparing two or more values that were taken over time or on different conditions, usually on small data sets. The bars can be horizontal lines or it can also be used to mass a point of view.
Descriptive statistic	Descriptive statistics are used to describe the main features of a collection of data in quantitative terms. Descriptive statistics are distinguished from inferential statistics (or inductive statistics), in that Descriptive statistics aim to quantitatively summarize a data set, rather than being used to support inferential statements about the population that the data are thought to represent. Even when a data analysis draws its main conclusions using inductive statistical analysis, Descriptive statistics are generally presented along with more formal analyses.
Eurobarometer	Eurobarometer is a series of surveys regularly performed on behalf of the European Commission since 1973. It produces reports of public opinion of certain issues relating to the European Union across the member states. The Eurobarometer results are published by the Public Opinion Analysis Sector of the European Commission Directorate-General Communication. The Eurobarometer program was initially launched and managed by Jacques-René Rabier, with the political support of both the European Parliament and Commission.
Opinion	An Opinion is a belief that may or may not be backed up with evidence, but which cannot be proved with that evidence. It is normally a subjective statement and may be the result of an emotion or an interpretation of facts; people may draw opposing Opinions from the same facts. There can be the public Opinion, or other types of Opinion.
Margin of error	The margin of error is a statistic expressing the amount of random sampling error in a survey's results. The larger the margin of error, the less faith one should have that the poll's reported results are close to the 'true' figures; that is, the figures for the whole population. Margin of error occurs whenever a population is incompletely sampled.
Parameter	Parameter can be interpreted in mathematics, logic, linguistics, environmental science and other disciplines.

	In its common meaning, the term is used to identify a characteristic, a feature, a measurable factor that can help in defining a particular system. It is an important element to take into consideration for the evaluation or for the comprehension of an event, a project or any situation.
ANOVA	In statistics, ANOVA is a collection of statistical models, and their associated procedures, in which the observed variance is partitioned into components due to different sources of variation. In its simplest form ANOVA provides a statistical test of whether or not the means of several groups are all equal, and therefore generalizes Student's two-sample t-test to more than two groups. ANOVAs are helpful because they possess a certain advantage over a two-sample t-test. Doing multiple two-sample t-tests would result in a largely increased chance of committing a type I error. For this reason, ANOVAs are useful in comparing three or more means. There are three conceptual classes of such models: · Fixed-effects models assume that the data came from normal populations which may differ only in their means. (Model 1) · Random effects models assume that the data describe a hierarchy of different populations whose differences are constrained by the hierarchy. (Model 2) · Mixed-effect models describe the situations where both fixed and random effects are present. (Model 3)
Random sampling	In random sampling every combination of items from the frame, or stratum, has a known probability of occurring, but these probabilities are not necessarily equal. With any form of sampling there is a risk that the sample may not adequately represent the population but with random sampling there is a large body of statistical theory which quantifies the risk and thus enables an appropriate sample size to be chosen.
Randomness	Randomness has somewhat disparate meanings as used in several different fields. It also has common meanings which may have loose connections with some of those more definite meanings. The Oxford English Dictionary defines 'random' thus:' Having no definite aim or purpose; not sent or guided in a particular direction; made, done, occurring, etc., without method or conscious choice; haphazard.'
Database	A Database is an integrated collection of logically related records or files consolidated into a common pool that provides data for one or multiple uses. One way of classifying Databases involves the type of content, for example: bibliographic, full-text, numeric, image. Other classification methods start from examining Database models architectures: see below.

Variance	In probability theory and statistics, the variance is a measure of how far a set of numbers is spread out. It is one of several descriptors of a probability distribution, describing how far the numbers lie from the mean (expected value). In particular, the variance is one of the moments of a distribution.

CHAPTER QUIZ: KEY TERMS, PEOPLE, PLACES, CONCEPTS

1. _____s

    In statistics, a _____ is a variable that can take on one of a limited, and usually fixed, number of possible values. _____s are often used to represent categorical data.

    A _____ that can take on exactly two values is termed a binary variable and is typically treated on its own as a special case.

    a. Categorical variable
    b. Compositional data
    c. Count data
    d. Cross-sectional data

2. An _____ is a belief that may or may not be backed up with evidence, but which cannot be proved with that evidence. It is normally a subjective statement and may be the result of an emotion or an interpretation of facts; people may draw opposing _____s from the same facts.

    There can be the public _____, or other types of _____.

    a. Internalism
    b. Opinion
    c. Animistic fallacy
    d. Extrinsic finality

3. . In statistics, a _____ is a type of table in a matrix format that displays the (multivariate) frequency distribution of the variables. It is often used to record and analyze the relation between two or more categorical variables.

    The term _____ was first used by Karl Pearson in 'On the Theory of Contingency and Its Relation to Association and Normal Correlation', part of the Drapers' Company Research Memoirs Biometric Series I published in 1904.

    a. Counternull
    b. Contingency table
    c. Cross tabulation

4. In probability theory and statistics, the _____ is a measure of how far a set of numbers is spread out. It is one of several descriptors of a probability distribution, describing how far the numbers lie from the mean (expected value). In particular, the _____ is one of the moments of a distribution.

   a. Variation ratio
   b. Variogram
   c. Variance
   d. Full width at half maximum

5. In statistics, _____ is the process of drawing conclusions from data subject to random variation, for example, observational errors or sampling variation. More substantially, the terms _____, statistical induction and inferential statistics are used to describe systems of procedures that can be used to draw conclusions from datasets arising from systems affected by random variation, such as observational errors, random sampling, or random experimentation. Initial requirements of such a system of procedures for inference and induction are that the system should produce reasonable answers when applied to well-defined situations and that it should be general enough to be applied across a range of situations.

   a. Trial and error
   b. Two-stage model of free will
   c. Validity
   d. Statistical inference

1. a
2. b
3. b
4. c
5. d

You can take the complete Chapter Practice Test

for Chapter 1. Statistics: The Art and Science of Learning from Data
on all key terms, persons, places, and concepts.

Online 99 Cents

http://www.epub27.14.20551.1.cram101.com/

Use www.Cram101.com for all your study needs

including Cram101's online interactive problem solving labs in

chemistry, statistics, mathematics, and more.

CHAPTER OUTLINE: KEY TERMS, PEOPLE, PLACES, CONCEPTS

	ANOVA
	Prevalence
	Sample survey
	Categorical variable
	Discrete probability distributions
	Pareto chart
	Relative frequency
	Bar graph
	Pie chart
	Pareto principle
	Histogram
	Probability
	Probability distribution
	Bimodal
	Bimodal distribution
	Symmetric distribution
	Unimodal
	Time series
	Median

Binomial

Binomial distribution

Outlier

Pollution

Statistic

Variance

Changing-criterion research design

Deviation

Standard deviation

Sum of squares

Ideal number

Population standard deviation

Empirical

Normal distribution

Parameter

Residual standard deviation

Sample mean

Sampling distribution

Percentile

_____ | Position

_____ | Quartile

_____ | Interquartile range

_____ | Box plot

_____ | Five-number summary

_____ | Potential outlier

_____ | Data analysis

_____ | Exploratory data analysis

_____ | Descriptive statistic

CHAPTER HIGHLIGHTS & NOTES: KEY TERMS, PEOPLE, PLACES, CONCEPTS

ANOVA

In statistics, ANOVA is a collection of statistical models, and their associated procedures, in which the observed variance is partitioned into components due to different sources of variation. In its simplest form ANOVA provides a statistical test of whether or not the means of several groups are all equal, and therefore generalizes Student's two-sample t-test to more than two groups. ANOVAs are helpful because they possess a certain advantage over a two-sample t-test. Doing multiple two-sample t-tests would result in a largely increased chance of committing a type I error. For this reason, ANOVAs are useful in comparing three or more means.

There are three conceptual classes of such models:

· Fixed-effects models assume that the data came from normal populations which may differ only in their means. (Model 1) · Random effects models assume that the data describe a hierarchy of different populations whose differences are constrained by the hierarchy. (Model 2) · Mixed-effect models describe the situations where both fixed and random effects are present.

Chapter 2. Exploring Data with Graphs and Numerical Summaries

Prevalence	In epidemiology, the prevalence of a health-related state (typically disease, but also other things like smoking or seatbelt use) in a statistical population is defined as the total number of cases of the risk factor in the population at a given time, or the total number of cases in the population, divided by the number of individuals in the population. It is used as an estimate of how common a disease is within a population over a certain period of time. It helps physicians or other health professionals understand the probability of certain diagnoses and is routinely used by epidemiologists, health care providers, government agencies and insurers.
Sample survey	Sample survey is a survey of a population made by using only a portion of the population.
Categorical variable	Categorical Variables In statistics, a categorical variable is a variable that can take on one of a limited, and usually fixed, number of possible values. Categorical variables are often used to represent categorical data. A categorical variable that can take on exactly two values is termed a binary variable and is typically treated on its own as a special case.
Discrete probability distributions	Discrete probability distributions arise in the mathematical description of probabilistic and statistical problems in which the values that might be observed are restricted to being within a pre-defined list of possible values. This list has either a finite number of members, or at most is countable. In probability theory, a probability distribution is called discrete if it is characterized by a probability mass function.
Pareto chart	A Pareto chart, is a type of chart that contains both bars and a line graph, where individual values are represented in descending order by bars, and the cumulative total is represented by the line. The left vertical axis is the frequency of occurrence, but it can alternatively represent cost or another important unit of measure. The right vertical axis is the cumulative percentage of the total number of occurrences, total cost, or total of the particular unit of measure.
Relative frequency	In a series of observations, or trials, the relative frequency of occurrence of an event E is calculated as the number of times the event E happened over the total number of observations made. The relative frequency density of occurrence of an event is the relative frequency of E divided by the size of the bin used to classify E.

Bar graph	A bar chart or Bar graph is a chart with rectangular bars with lengths proportional to the values that they represent. Bar charts are used for comparing two or more values that were taken over time or on different conditions, usually on small data sets. The bars can be horizontal lines or it can also be used to mass a point of view.
Pie chart	A pie chart is a circular chart divided into sectors, illustrating proportion. In a pie chart, the arc length of each sector (and consequently its central angle and area), is proportional to the quantity it represents. When angles are measured with 1 turn as unit then a number of percent is identified with the same number of centiturns.
Pareto principle	The Pareto principle states that, for many events, roughly 80% of the effects come from 20% of the causes. Business-management consultant Joseph M. Juran suggested the principle and named it after Italian economist Vilfredo Pareto, who observed in 1906 that 80% of the land in Italy was owned by 20% of the population; he developed the principle by observing that 20% of the pea pods in his garden contained 80% of the peas. It is a common rule of thumb in business; e.g., '80% of your sales come from 20% of your clients'.
Histogram	In statistics, a histogram is a graphical representation showing a visual impression of the distribution of data. It is an estimate of the probability distribution of a continuous variable and was first introduced by Karl Pearson. A histogram consists of tabular frequencies, shown as adjacent rectangles, erected over discrete intervals (bins), with an area equal to the frequency of the observations in the interval.
Probability	Probability is a way of expressing knowledge or belief that an event will occur or has occurred. The concept has been given an exact mathematical meaning in probability theory, which is used extensively in such areas of study as mathematics, statistics, finance, gambling, science, Artificial intelligence/Machine learning and philosophy to draw conclusions about the likelihood of potential events and the underlying mechanics of complex systems. Interpretations The word probability does not have a consistent direct definition.
Probability distribution	In probability theory, a probability mass, probability density, or probability distribution is a function that describes the probability of a random variable taking certain values.

	For a more precise definition one needs to distinguish between discrete and continuous random variables. In the discrete case, one can easily assign a probability to each possible value: when throwing a die, each of the six values 1 to 6 has the probability 1/6. In contrast, when a random variable takes values from a continuum, probabilities are nonzero only if they refer to finite intervals: in quality control one might demand that the probability of a '500 g' package containing between 490 g and 510 g should be no less than 98%.
Bimodal	In statistics, a Bimodal distribution is a continuous probability distribution with two different modes. These appear as distinct peaks (local maxima) in the probability density function. Examples of variables with Bimodal distributions include the time between eruptions of certain geysers, the color of galaxies, the size of worker weaver ants, the age of incidence of Hodgkin's lymphoma, the speed of inactivation of the drug isoniazid in US adults, and the absolute magnitude of novae. A Bimodal distribution most commonly arises as a mixture of two different unimodal distributions (i.e. distributions having only one mode).
Bimodal distribution	In statistics, a bimodal distribution is a continuous probability distribution with two different modes. These appear as distinct peaks (local maxima) in the probability density function. Examples of variables with bimodal distributions include the time between eruptions of certain geysers, the color of galaxies, the size of worker weaver ants, the age of incidence of Hodgkin's lymphoma, the speed of inactivation of the drug isoniazid in US adults, the absolute magnitude of novae, and the circadian activity patterns of those crepuscular animals that are active both in morning and evening twilight.
Symmetric distribution	Symmetric distribution is distribution without skewness with the opposing sides symmetric about the mean and media.
Unimodal	In mathematics, a function f(x) between two ordered sets is unimodal if for some value m (the mode), it is monotonically increasing for x ≤ m and monotonically decreasing for x ≥ m. In that case, the maximum value of f(x) is f(m) and there are no other local maxima. Examples of unimodal functions: · Quadratic polynomial with a negative quadratic coefficient · Logistic map · Tent map Function $f(x)$ is 'S-unimodal' if its Schwartzian derivative is negative for all $x \neq 0$.

Time series	In statistics, signal processing, econometrics and mathematical finance, a time series is a sequence of data points, measured typically at successive time instants spaced at uniform time intervals. Examples of time series are the daily closing value of the Dow Jones index or the annual flow volume of the Nile River at Aswan. Time series analysis comprises methods for analyzing time series data in order to extract meaningful statistics and other characteristics of the data.
Median	In probability theory and statistics, a median is described as the numerical value separating the higher half of a sample, a population, or a probability distribution, from the lower half. The median of a finite list of numbers can be found by arranging all the observations from lowest value to highest value and picking the middle one. If there is an even number of observations, then there is no single middle value; the median is then usually defined to be the mean of the two middle values.
Binomial	In elementary algebra, a Binomial is a polynomial with two terms--the sum of two monomials--often bound by parenthesis or brackets when operated upon. It is the simplest kind of polynomial other than monomials.

· The Binomial $a^2 - b^2$ can be factored as the product of two other Binomials:

$a^2 - b^2 = (a + b)(a - b)$.

This is a special case of the more general formula:

$$a^{n+1} - b^{n+1} = (a - b) \sum_{k=0}^{n} a^k b^{n-k}$$.

· The product of a pair of linear Binomials $(ax + b)$ and $(cx + d)$ is:

$(ax + b)(cx + d) = acx^2 + axd + bcx + bd$.

· A Binomial raised to the nth power, represented as

$(a + b)^n$

	can be expanded by means of the Binomial theorem or, equivalently, using Pascal's triangle. Taking a simple example, the perfect square Binomial $(p + q)^2$ can be found by squaring the first term, adding twice the product of the first and second terms and finally adding the square of the second term, to give $p^2 + 2pq + q^2$.
	· A simple but interesting application of the cited Binomial formula is the '(m,n)-formula' for generating Pythagorean triples: for m < n, let $a = n^2 - m^2$, $b = 2mn$, $c = n^2 + m^2$, then $a^2 + b^2 = c^2$.
Binomial distribution	In probability theory and statistics, the binomial distribution is the discrete probability distribution of the number of successes in a sequence of n independent yes/no experiments, each of which yields success with probability p. Such a success/failure experiment is also called a Bernoulli experiment or Bernoulli trial; when n = 1, the binomial distribution is a Bernoulli distribution. The binomial distribution is the basis for the popular binomial test of statistical significance.
Outlier	In statistics, an outlier is an observation that is numerically distant from the rest of the data. Grubbs defined an outlier as:'
	An outlying observation, or outlier, is one that appears to deviate markedly from other members of the sample in which it occurs. '
	Outliers can occur by chance in any distribution, but they are often indicative either of measurement error or that the population has a heavy-tailed distribution.
Pollution	Pollution is the introduction of contaminants into an environment that causes instability, disorder, harm or discomfort to the ecosystem i.e. physical systems or living organisms . pollution can take the form of chemical substances, or energy, such as noise, heat, or light energy. Pollutants, the elements of pollution, can be foreign substances or energies, or naturally occurring; when naturally occurring, they are considered contaminants when they exceed natural levels.
Statistic	A statistic is a single measure of some attribute of a sample (e.g. its arithmetic mean value). It is calculated by applying a function (statistical algorithm) to the values of the items comprising the sample which are known together as a set of data.
	More formally, statistical theory defines a statistic as a function of a sample where the function itself is independent of the sample's distribution; that is, the function can be stated before realisation of the data.

Variance	In probability theory and statistics, the variance is a measure of how far a set of numbers is spread out. It is one of several descriptors of a probability distribution, describing how far the numbers lie from the mean (expected value). In particular, the variance is one of the moments of a distribution.
Changing-criterion research design	In a changing-criterion research design a criterion for reinforcement is changed across the experiment to demonstrate the functional relationship between the reinforcement and the behavior. See Mark Dixon's work with a participant using a short video clip to generate a preference for a progressively delayed variable reinforcement over a fixed shorter delay reinforcement in physical therapy.
Deviation	In mathematics and statistics, deviation is a measure of difference between the observed value and the mean. The sign of deviation (positive or negative), reports the direction of that difference (it is larger when the sign is positive, and smaller if it is negative). The magnitude of the value indicates the size of the difference.
Standard deviation	Standard deviation is a widely used measurement of variability or diversity used in statistics and probability theory. It shows how much variation or 'dispersion' there is from the 'average' (mean, or expected/budgeted value). A low standard deviation indicates that the data points tend to be very close to the mean, whereas high standard deviation indicates that the data are spread out over a large range of values.
Sum of squares	Sum of squares is a concept that permeates much of inferential statistics and descriptive statistics. More properly, it is the sum of squared deviations. Mathematically, it is an unscaled, or unadjusted measure of dispersion (also called variability).
Ideal number	In number theory an ideal number is an algebraic integer which represents an ideal in the ring of integers of a number field; the idea was developed by Ernst Kummer, and led to Richard Dedekind's definition of ideals for rings. An ideal in the ring of integers of an algebraic number field is principal if it consists of multiples of a single element of the ring, and nonprincipal otherwise. By the principal ideal theorem any nonprincipal ideal becomes principal when extended to an ideal of the Hilbert class field.
Population standard deviation	Population standard deviation is the amountof variation one would expect to see in a population, for some given attribute. To determine the actual standard deviation for a population, you would have to sample each individual member of a population for the specifictrait you are investigating.
Empirical	The word empirical denotes information acquired by means of observation or experimentation. Empirical data are data produced by an observation or experiment.

Chapter 2. Exploring Data with Graphs and Numerical Summaries

Normal distribution	In probability theory, the normal (or Gaussian) distribution is a continuous probability distribution that has a bell-shaped probability density function, known as the Gaussian function or informally the bell curve: $$f(x; \mu, \sigma^2) = \frac{1}{\sigma\sqrt{2\pi}} e^{-\frac{1}{2}\left(\frac{x-\mu}{\sigma}\right)^2}$$ The parameter μ is the mean or expectation (location of the peak) and $\sigma^{?2}$ is the variance. σ is known as the standard deviation. The distribution with μ = 0 and $\sigma^{?2}$ = 1 is called the standard normal distribution or the unit normal distribution.
Parameter	Parameter can be interpreted in mathematics, logic, linguistics, environmental science and other disciplines. In its common meaning, the term is used to identify a characteristic, a feature, a measurable factor that can help in defining a particular system. It is an important element to take into consideration for the evaluation or for the comprehension of an event, a project or any situation.
Residual standard deviation	The residual standard deviation is a goodness-fit measure. The smaller the residual standard deviation, the closer is the fit to the data.
Sample mean	The sample mean or empirical mean and the sample covariance are statistics computed from a collection of data, thought of as being random. Given a random sample $\mathbf{x}_1, \dots, \mathbf{x}_N$ from an n-dimensional random variable $\mathbf{X}$ (i.e., realizations of N independent random variables with the same distribution as $\mathbf{X}$), the sample mean is $$\bar{\mathbf{x}} = \frac{1}{N}\sum_{k=1}^{N} \mathbf{x}_k.$$ In coordinates, writing the vectors as columns, $$\mathbf{x}_k = \begin{bmatrix} x_{1k} \\ \vdots \\ x_{nk} \end{bmatrix}, \quad \bar{\mathbf{x}} = \begin{bmatrix} \bar{x}_1 \\ \vdots \\ \bar{x}_n \end{bmatrix},$$ the entries of the sample mean are

$$\bar{x}_i = \frac{1}{N}\sum_{k=1}^{N} x_{ik}, \quad i = 1, \ldots, n.$$

The sample covariance of $\mathbf{x}_1, \ldots, \mathbf{x}_N$ is the n-by-n matrix $\mathbf{Q} = [q_{ij}]$ with the entries given by

$$q_{ij} = \frac{1}{N-1}\sum_{k=1}^{N} (x_{ik} - \bar{x}_i)(x_{jk} - \bar{x}_j)$$

The sample mean and the sample covariance matrix are unbiased estimates of the mean and the covariance matrix of the random variable $\mathbf{X}$. The reason why the sample covariance matrix has $N-1$ in the denominator rather than N is essentially that the population mean E(X) is not known and is replaced by the sample mean $\bar{x}$.

Sampling distribution	In statistics, a sampling distribution is the probability distribution of a given statistic based on a random sample. Sampling distributions are important in statistics because they provide a major simplification on the route to statistical inference. More specifically, they allow analytical considerations to be based on the sampling distribution of a statistic, rather than on the joint probability distribution of all the individual sample values.
Percentile	In statistics and the social sciences, a percentile is the value of a variable below which a certain percent of observations fall. For example, the 20th percentile is the value (or score) below which 20 percent of the observations may be found. The term percentile and the related term percentile rank are often used in the reporting of scores from norm-referenced tests.
Position	In geometry, a position, location, or radius vector, usually denoted $\mathbf{r}$, is a vector which represents the position of a point P in space in relation to an arbitrary reference origin O. It corresponds to the displacement from O to P: $\mathbf{r} = \overrightarrow{OP}$.

The concept typically applies to two- or three-dimensional space, but can be easily generalized to Euclidean spaces with a higher number of dimensions. Applications •In linear algebra, a position vector can be expressed as a linear combination of basis vectors.•The kinematic movement of a point mass can be described by a vector-valued function giving the position $\mathbf{r}(t)$ as a function of the scalar time parameter t. These are used in mechanics and dynamics to keep track of the positions of particles, point masses, or rigid objects.•In differential geometry, position vector fields are used to describe continuous and differentiable space curves, in which case the independent parameter needs not be time, but can be (e.g). |

Chapter 2. Exploring Data with Graphs and Numerical Summaries

Quartile	In descriptive statistics, a Quartile is any of the three values which divide the sorted data set into four equal parts, so that each part represents one fourth of the sampled population.
	In epidemiology, the Quartiles are the four ranges defined by the three values discussed here.
	· first Quartile (designated Q_1) = lower Quartile = cuts off lowest 25% of data = 25th percentile · second Quartile (designated Q_2) = median = cuts data set in half = 50th percentile · third Quartile (designated Q_3) = upper Quartile = cuts off highest 25% of data, or lowest 75% = 75th percentile
	The difference between the upper and lower Quartiles is called the interQuartile range.
	There is no universal agreement on choosing the Quartile values.
	The formula for locating the position of the observation at a given percentile, y, with n data points sorted in ascending order is:
	$$L_y = (n)\left(\frac{y}{100}\right)$$
	· Case 1: If L is a whole number, then the value will be found halfway between positions L and L+1 · Case 2: If L is a decimal, round up to the nearest whole number. (for example, L = 1.2 becomes 2) Example 4. Boxplot (with Quartiles and an interQuartile range) and a probability density function (pdf) of a normal $N(0,1\sigma^2)$ population
	One possible rule (employed by the TI-83 calculator boxplot and 1-Var Stats functions) is as follows:
	· Use the median to divide the ordered data set into two halves.
Interquartile range	In descriptive statistics, the interquartile range also called the midspread or middle fifty, is a measure of statistical dispersion, being equal to the difference between the upper and lower quartiles. $IQR = Q_3 - Q_1$
	Unlike (total) range, the interquartile range is a robust statistic, having a breakdown point of 25%, and is thus often preferred to the total range.
	The IQR is used to build box plots, simple graphical representations of a probability distribution.

Box plot	In descriptive statistics, a box plot is a convenient way of graphically depicting groups of numerical data through their five-number summaries: the smallest observation (sample minimum), lower quartile (Q1), median (Q2), upper quartile (Q3), and largest observation (sample maximum). A boxplot may also indicate which observations, if any, might be considered outliers.
	Boxplots display differences between populations without making any assumptions of the underlying statistical distribution: they are non-parametric.
Five-number summary	The five-number summary is a descriptive statistic that provides information about a set of observations. It consists of the five most important sample percentiles:•the sample minimum (smallest observation)•the lower quartile or first quartile•the median (middle value)•the upper quartile or third quartile•the sample maximum (largest observation)
	In order for these statistics to exist the observations must be from a univariate variable that can be measured on an ordinal, interval or ratio scale. Use and representation
	The five-number summary provides a concise summary of the distribution of the observations.
Potential outlier	A potential outlier should be examined to see if they are possibly erroneous. If the data point is in error, it should be corrected if possible and deleted if it is not possible. If there is no reason to believe that the outlying point is in error, it should not be deleted without careful consideration. However, the use of more robust techniques may be warranted.
Data analysis	Analysis of data is a process of inspecting, cleaning, transforming, and modeling data with the goal of highlighting useful information, suggesting conclusions, and supporting decision making. Data analysis has multiple facets and approaches, encompassing diverse techniques under a variety of names, in different business, science, and social science domains.
	Data mining is a particular data analysis technique that focuses on modeling and knowledge discovery for predictive rather than purely descriptive purposes.
Exploratory data analysis	In statistics, exploratory data analysis is an approach to analyzing data sets to summarize their main characteristics in easy-to-understand form, often with visual graphs, without using a statistical model or having formulated a hypothesis. Exploratory data analysis was promoted by John Tukey to encourage statisticians visually to examine their data sets, to formulate hypotheses that could be tested on new data-sets.
	Tukey's championing of EDA encouraged the development of statistical computing packages, especially S at Bell Labs: The S programming language inspired the systems 'S'-PLUS and R.

Chapter 2. Exploring Data with Graphs and Numerical Summaries

Descriptive statistic	Descriptive statistics are used to describe the main features of a collection of data in quantitative terms. Descriptive statistics are distinguished from inferential statistics (or inductive statistics), in that Descriptive statistics aim to quantitatively summarize a data set, rather than being used to support inferential statements about the population that the data are thought to represent. Even when a data analysis draws its main conclusions using inductive statistical analysis, Descriptive statistics are generally presented along with more formal analyses.

1. In statistics, _____ is a collection of statistical models, and their associated procedures, in which the observed variance is partitioned into components due to different sources of variation. In its simplest form _____ provides a statistical test of whether or not the means of several groups are all equal, and therefore generalizes Student's two-sample t-test to more than two groups. _____s are helpful because they possess a certain advantage over a two-sample t-test. Doing multiple two-sample t-tests would result in a largely increased chance of committing a type I error. For this reason, _____s are useful in comparing three or more means.

There are three conceptual classes of such models:

· Fixed-effects models assume that the data came from normal populations which may differ only in their means. (Model 1) · Random effects models assume that the data describe a hierarchy of different populations whose differences are constrained by the hierarchy. (Model 2) · Mixed-effect models describe the situations where both fixed and random effects are present. (Model 3)

    a. Analysis of variance
    b. Acceptable quality level
    c. ANOVA
    d. initial condition

2. In statistics, signal processing, econometrics and mathematical finance, a _____ is a sequence of data points, measured typically at successive time instants spaced at uniform time intervals. Examples of _____ are the daily closing value of the Dow Jones index or the annual flow volume of the Nile River at Aswan. _____ analysis comprises methods for analyzing _____ data in order to extract meaningful statistics and other characteristics of the data.

    a. Time series
    b. Decomposition of time series
    c. Detrended fluctuation analysis
    d. Fourier analysis

3. . In statistics, a _____ is a graphical representation showing a visual impression of the distribution of data.

It is an estimate of the probability distribution of a continuous variable and was first introduced by Karl Pearson. A _____ consists of tabular frequencies, shown as adjacent rectangles, erected over discrete intervals (bins), with an area equal to the frequency of the observations in the interval.

a. Log-log plot
b. Lorenz curve
c. Histogram
d. Radar chart

4. The _____ is a descriptive statistic that provides information about a set of observations. It consists of the five most important sample percentiles:•the sample minimum (smallest observation)•the lower quartile or first quartile•the median (middle value)•the upper quartile or third quartile•the sample maximum (largest observation)

In order for these statistics to exist the observations must be from a univariate variable that can be measured on an ordinal, interval or ratio scale. Use and representation

The _____ provides a concise summary of the distribution of the observations.

a. Frequency
b. Frequency distribution
c. Generalized entropy index
d. Five-number summary

5. In a series of observations, or trials, the _____ of occurrence of an event E is calculated as the number of times the event E happened over the total number of observations made. The _____ density of occurrence of an event is the _____ of E divided by the size of the bin used to classify E.

a. Box plot
b. Nonproportional quota sampling
c. Best-fit line
d. Relative frequency

1. c
2. a
3. c
4. d
5. d

You can take the complete Chapter Practice Test

for Chapter 2. Exploring Data with Graphs and Numerical Summaries
on all key terms, persons, places, and concepts.

Online 99 Cents

http://www.epub27.14.20551.2.cram101.com/

Use www.Cram101.com for all your study needs

including Cram101's online interactive problem solving labs in

chemistry, statistics, mathematics, and more.

Chapter 3. Association: Contingency, Correlation, and Regression

	Response variable
	Categorical variable
	Organic food
	Cell
	Contingency table
	Residue
	Parameter
	Independence
	Probability
	Statistic
	Internet
	Positive number
	Election
	Correlation
	Absolute value
	Paradox
	Quadrant
	Regression line
	Y-intercept

Regression equation

Standard error

Least squares

Squared

Extrapolation

Forecasting

Outlier

Potential outlier

Higher

Confounding

Chapter 3. Association: Contingency, Correlation, and Regression

Response variable	The terms 'dependent variable' and 'independent variable' are used in similar but subtly different ways in mathematics and statistics as part of the standard terminology in those subjects. They are used to distinguish between two types of quantities being considered, separating them into those available at the start of a process and those being created by it, where the latter (dependent variables) are dependent on the former (independent variables).
	The independent variable is typically the variable representing the value being manipulated or changed and the dependent variable is the observed result of the independent variable being manipulated. For example concerning nutrition, the independent variable of daily vitamin C intake (how much vitamin C one consumes) can influence the dependent variable of life expectancy (the average age one attains). Over some period of time, scientists will control the vitamin C intake in a substantial group of people. One part of the group will be given a daily high dose of vitamin C, and the remainder will be given a placebo pill (so that they are unaware of not belonging to the first group) without vitamin C. The scientists will investigate if there is any statistically significant difference in the life span of the people who took the high dose and those who took the placebo (no dose). The goal is to see if the independent variable of high vitamin C dosage has a correlation with the dependent variable of people's life span. The designation independent/dependent is clear in this case, because if a correlation is found, it cannot be that life span has influenced vitamin C intake, but an influence in the other direction is possible. Use in mathematics
	In calculus, a function is a map whose action is specified on variables. Take x and y to be two variables. A function f may map x to some expression in x. Assigning $y = f(x)$ gives a relation between x and y. If there is some relation specifying y in terms of x, then y is known as a 'dependent variable' (and x is an 'independent variable'). Use in statistics Controlled experiments
	In a statistics experiment, the dependent variable is the event studied and expected to change whenever the independent variable is altered.
	In the design of experiments, an independent variable's values are controlled or selected by the experimenter to determine its relationship to an observed phenomenon (i.e., the dependent variable). In such an experiment, an attempt is made to find evidence that the values of the independent variable determine the values of the dependent variable. The independent variable can be changed as required, and its values do not represent a problem requiring explanation in an analysis, but are taken simply as given. The dependent variable, on the other hand, usually cannot be directly controlled.
	Controlled variables are also important to identify in experiments. They are the variables that are kept constant to prevent their influence on the effect of the independent variable on the dependent.

Every experiment has a controlling variable, and it is necessary to not change it, or the results of the experiment won't be valid.

'Extraneous variables' are those that might affect the relationship between the independent and dependent variables. Extraneous variables are usually not theoretically interesting. They are measured in order for the experimenter to compensate for them. For example, an experimenter who wishes to measure the degree to which caffeine intake (the independent variable) influences explicit recall for a word list (the dependent variable) might also measure the participant's age (extraneous variable). She can then use these age data to control for the uninteresting effect of age, clarifying the relationship between caffeine and memory.

In summary:•Independent variables answer the question 'What do I change?'•Dependent variables answer the question 'What do I observe?'•Controlled variables answer the question 'What do I keep the same?'•Extraneous variables answer the question 'What uninteresting variables might mediate the effect of the IV on the DV?'Alternative terminology in statistics

In statistics, the dependent/independent variable terminology is used more widely than just in relation to controlled experiments. For example the data analysis of two jointly varying quantities may involve treating each in turn as the dependent variable and the other as the independent variable. However, for general usage, the pair response variable and explanatory variable is preferable as quantities treated as 'independent variables' are rarely statistically independent.

Categorical variable	Categorical Variables

In statistics, a categorical variable is a variable that can take on one of a limited, and usually fixed, number of possible values. Categorical variables are often used to represent categorical data.

A categorical variable that can take on exactly two values is termed a binary variable and is typically treated on its own as a special case.

Organic food	Organic foods are foods that are produced using methods that do not involve modern synthetic

inputs such as synthetic pesticides and chemical fertilizers, do not contain genetically modified organisms, and are not processed using irradiation, industrial solvents, or chemical food additives.

The weight of the available scientific evidence has not shown a significant difference between organic and more conventionally grown food in terms of safety, nutritional value, or taste.

Chapter 3. Association: Contingency, Correlation, and Regression

Cell	In geometry, a cell is a three-dimensional element that is part of a higher-dimensional object.
	In polytopes
	A cell is a three-dimensional polyhedron element that is part of the boundary of a higher-dimensional polytope, such as a polychoron (4-polytope) or honeycomb (3-space tessellation).
	For example, a cubic honeycomb is made of cubic cells, with 4 cubes on each edge.
Contingency table	In statistics, a contingency table is a type of table in a matrix format that displays the (multivariate) frequency distribution of the variables. It is often used to record and analyze the relation between two or more categorical variables.
	The term contingency table was first used by Karl Pearson in 'On the Theory of Contingency and Its Relation to Association and Normal Correlation', part of the Drapers' Company Research Memoirs Biometric Series I published in 1904.
Residue	In mathematics, more specifically complex analysis, the residue is a complex number equal to the contour integral of a meromorphic function along a path enclosing one of its singularities. Residues can be computed quite easily and, once known, allow the determination of general contour integrals via the residue theorem.
	The residue of a meromorphic function f at an isolated singularity a, often denoted $\mathrm{Res}(f, a)$ is the unique value R such that f(z) − R / (z − a) has an analytic antiderivative in a punctured disk $0 < |z - a| < \delta$.
Parameter	Parameter can be interpreted in mathematics, logic, linguistics, environmental science and other disciplines.
	In its common meaning, the term is used to identify a characteristic, a feature, a measurable factor that can help in defining a particular system. It is an important element to take into consideration for the evaluation or for the comprehension of an event, a project or any situation.
Independence	In probability theory, to say that two events are independent intuitively means that the occurrence of one event makes it neither more nor less probable that the other occurs.

For example:•The event of getting a 6 the first time a die is rolled and the event of getting a 6 the second time are independent.•By contrast, the event of getting a 6 the first time a die is rolled and the event that the sum of the numbers seen on the first and second trials is 8 are not independent.•If two cards are drawn with replacement from a deck of cards, the event of drawing a red card on the first trial and that of drawing a red card on the second trial are independent.•By contrast, if two cards are drawn without replacement from a deck of cards, the event of drawing a red card on the first trial and that of drawing a red card on the second trial are again not independent.

Similarly, two random variables are independent if the conditional probability distribution of either given the observed value of the other is the same as if the other's value had not been observed. The concept of independence extends to dealing with collections of more than two events or random variables.

| Probability | Probability is a way of expressing knowledge or belief that an event will occur or has occurred. The concept has been given an exact mathematical meaning in probability theory, which is used extensively in such areas of study as mathematics, statistics, finance, gambling, science, Artificial intelligence/Machine learning and philosophy to draw conclusions about the likelihood of potential events and the underlying mechanics of complex systems.

Interpretations

The word probability does not have a consistent direct definition. |

Statistic | A statistic is a single measure of some attribute of a sample (e.g. its arithmetic mean value). It is calculated by applying a function (statistical algorithm) to the values of the items comprising the sample which are known together as a set of data.

More formally, statistical theory defines a statistic as a function of a sample where the function itself is independent of the sample's distribution; that is, the function can be stated before realisation of the data.

Internet | The Internet is a global system of interconnected computer networks that use the standard Internet Protocol Suite (TCP/IP) to serve billions of users worldwide. It is a network of networks that consists of millions of private, public, academic, business, and government networks, of local to global scope, that are linked by a broad array of electronic and optical networking technologies. The Internet carries a vast range of information resources and services, such as the inter-linked hypertext documents of the World Wide Web (WWW) and the infrastructure to support electronic mail.

Chapter 3. Association: Contingency, Correlation, and Regression

Positive number	Being a positive number is a property of a number which is real, or a member of a subset of real numbers such as rational and integer numbers. A positive number is one that is greater than zero, such as $\sqrt{2}$, 1.414, 1. Zero itself is neither positive nor negative. The non-negative numbers are the numbers that are not negative .
Election	An election is a formal decision-making process by which a population chooses an individual to hold public office. Elections have been the usual mechanism by which modern representative democracy operates since the 17th century. Elections may fill offices in the legislature, sometimes in the executive and judiciary, and for regional and local government.
Correlation	In statistics, correlation (often measured as a correlation coefficient, ρ) indicates the strength and direction of a relationship between two random variables. The commonest use refers to a linear relationship. In general statistical usage, correlation or co-relation refers to the departure of two random variables from independence.
Absolute value	In mathematics, the absolute value a of a real number a is as numerical value without regard to its sign. So, for example, 3 is the absolute value of both 3 and −3. Generalizations of the absolute value for real numbers occur in a wide variety of mathematical settings.
Paradox	A Paradox is a true statement or group of statements that leads to a contradiction or a situation which defies intuition. The term is also used for an apparent contradiction that actually expresses a non-dual truth (cf. kÅan, Catuskoti).
Quadrant	Quadrant is a rectangular divisions on the celestial sphere that is used for constellation navigation. The celestial sphere is divided into northern hemisphere and southern hemisphere with each hemisphere divided into four six-hour arcs. It is symbolized by using the capital letter N or S, followed by Q for quadrant, and quadrant numbers 1, 2, 3, or 4. However to determine exact quadrants, the midpoints of the constellations must be used.
Regression line	Regression line is a line drawn through a scatterplot of two variables. The line is chosen so that it comes as close to the points as possible.
Y-intercept	In coordinate geometry, using the common convention that the horizontal axis represents a variable x and the vertical axis represents a variable y, a y-intercept is a point where the graph of a function or relation intersects with the y-axis of the coordinate system. As such, these points satisfy x=0.

Regression equation	The regression equation represents the relation between selected values of one variable (x) and observed values of the other (y); it permits the prediction of the most probable values of y.
Standard error	The standard error is the standard deviation of the sampling distribution of a statistic. The term may also be used to refer to an estimate of that standard deviation, derived from a particular sample used to compute the estimate. For example, the sample mean is the usual estimator of a population mean.
Least squares	The method of least squares is a standard approach to the approximate solution of overdetermined systems, i.e., sets of equations in which there are more equations than unknowns. 'Least squares' means that the overall solution minimizes the sum of the squares of the errors made in the results of every single equation. The most important application is in data fitting.
Squared	In algebra, the square of a number is that number multiplied by itself. To square a quantity is to multiply it by itself. Its notation is a superscripted '2'; a number x squared is written as x^2.
Extrapolation	In mathematics, extrapolation is the process of constructing new data points. It is similar to the process of interpolation, which constructs new points between known points, but the results of extrapolations are often less meaningful, and are subject to greater uncertainty. It may also mean extension of a method, assuming similar methods will be applicable.
Forecasting	Forecasting is the process of making statements about events whose actual outcomes (typically) have not yet been observed. A commonplace example might be estimation for some variable of interest at some specified future date. Prediction is a similar, but more general term.
Outlier	In statistics, an outlier is an observation that is numerically distant from the rest of the data. Grubbs defined an outlier as:' An outlying observation, or outlier, is one that appears to deviate markedly from other members of the sample in which it occurs. ' Outliers can occur by chance in any distribution, but they are often indicative either of measurement error or that the population has a heavy-tailed distribution.
Potential outlier	A potential outlier should be examined to see if they are possibly erroneous. If the data point is in error, it should be corrected if possible and deleted if it is not possible. If there is no reason to believe that the outlying point is in error, it should not be deleted without careful consideration.

Chapter 3. Association: Contingency, Correlation, and Regression

Higher	In Scotland the Higher (Scottish Gaelic: An Àrd Ìre) is one of the national school-leaving certificate exams and university entrance qualifications of the Scottish Qualifications Certificate (SQC) offered by the Scottish Qualifications Authority. It superseded the old Higher Grade on the Scottish Certificate of Education (SCE). Both are normally referred to simply as 'Highers'.
Confounding	In statistics, a confounding variable (also confounding factor, lurking variable, a confound, or confounder) is an extraneous variable in a statistical model that correlates (positively or negatively) with both the dependent variable and the independent variable. The methodologies of scientific studies therefore need to control for these factors to avoid a false positive (Type I) error; an erroneous conclusion that the dependent variables are in a causal relationship with the independent variable. Such a relation between two observed variables is termed a spurious relationship.

1. . The terms 'dependent variable' and 'independent variable' are used in similar but subtly different ways in mathematics and statistics as part of the standard terminology in those subjects. They are used to distinguish between two types of quantities being considered, separating them into those available at the start of a process and those being created by it, where the latter (dependent variables) are dependent on the former (independent variables).

The independent variable is typically the variable representing the value being manipulated or changed and the dependent variable is the observed result of the independent variable being manipulated. For example concerning nutrition, the independent variable of daily vitamin C intake (how much vitamin C one consumes) can influence the dependent variable of life expectancy (the average age one attains). Over some period of time, scientists will control the vitamin C intake in a substantial group of people. One part of the group will be given a daily high dose of vitamin C, and the remainder will be given a placebo pill (so that they are unaware of not belonging to the first group) without vitamin C. The scientists will investigate if there is any statistically significant difference in the life span of the people who took the high dose and those who took the placebo (no dose). The goal is to see if the independent variable of high vitamin C dosage has a correlation with the dependent variable of people's life span. The designation independent/dependent is clear in this case, because if a correlation is found, it cannot be that life span has influenced vitamin C intake, but an influence in the other direction is possible. Use in mathematics

In calculus, a function is a map whose action is specified on variables. Take x and y to be two variables. A function f may map x to some expression in x. Assigning $y = f(x)$ gives a relation between x and y. If there is some relation specifying y in terms of x, then y is known as a 'dependent variable' (and x is an 'independent variable'). Use in statistics Controlled experiments

In a statistics experiment, the dependent variable is the event studied and expected to change whenever the independent variable is altered.

In the design of experiments, an independent variable's values are controlled or selected by the experimenter to determine its relationship to an observed phenomenon (i.e., the dependent variable). In such an experiment, an attempt is made to find evidence that the values of the independent variable determine the values of the dependent variable. The independent variable can be changed as required, and its values do not represent a problem requiring explanation in an analysis, but are taken simply as given. The dependent variable, on the other hand, usually cannot be directly controlled.

Controlled variables are also important to identify in experiments. They are the variables that are kept constant to prevent their influence on the effect of the independent variable on the dependent. Every experiment has a controlling variable, and it is necessary to not change it, or the results of the experiment won't be valid.

'Extraneous variables' are those that might affect the relationship between the independent and dependent variables. Extraneous variables are usually not theoretically interesting. They are measured in order for the experimenter to compensate for them. For example, an experimenter who wishes to measure the degree to which caffeine intake (the independent variable) influences explicit recall for a word list (the dependent variable) might also measure the participant's age (extraneous variable). She can then use these age data to control for the uninteresting effect of age, clarifying the relationship between caffeine and memory.

In summary:•Independent variables answer the question 'What do I change?'•Dependent variables answer the question 'What do I observe?'•Controlled variables answer the question 'What do I keep the same?'•Extraneous variables answer the question 'What uninteresting variables might mediate the effect of the IV on the DV?'Alternative terminology in statistics

In statistics, the dependent/independent variable terminology is used more widely than just in relation to controlled experiments. For example the data analysis of two jointly varying quantities may involve treating each in turn as the dependent variable and the other as the independent variable. However, for general usage, the pair _____ and explanatory variable is preferable as quantities treated as 'independent variables' are rarely statistically independent.

a. Ridge regression
b. Robust measures of scale
c. Response variable
d. Scale parameter

2. An _____ is a formal decision-making process by which a population chooses an individual to hold public office. _____s have been the usual mechanism by which modern representative democracy operates since the 17th century. _____s may fill offices in the legislature, sometimes in the executive and judiciary, and for regional and local government.

a. Ellsberg paradox
b. IDF model
c. Unanimity
d. Election

3. . In statistics, _____ (often measured as a _____ coefficient, ρ) indicates the strength and direction of a relationship between two random variables. The commonest use refers to a linear relationship. In general statistical usage, _____ or co-relation refers to the departure of two random variables from independence.

a. Covariance matrix
b. Correlation
c. Sample covariance
d. Sample covariance matrix

4. In probability theory, to say that two events are independent intuitively means that the occurrence of one event makes it neither more nor less probable that the other occurs. For example:•The event of getting a 6 the first time a die is rolled and the event of getting a 6 the second time are independent.•By contrast, the event of getting a 6 the first time a die is rolled and the event that the sum of the numbers seen on the first and second trials is 8 are not independent.•If two cards are drawn with replacement from a deck of cards, the event of drawing a red card on the first trial and that of drawing a red card on the second trial are independent.•By contrast, if two cards are drawn without replacement from a deck of cards, the event of drawing a red card on the first trial and that of drawing a red card on the second trial are again not independent.

Similarly, two random variables are independent if the conditional probability distribution of either given the observed value of the other is the same as if the other's value had not been observed. The concept of _____ extends to dealing with collections of more than two events or random variables.

a. Abu Mansur Abd al-Qahir ibn Tahir ibn Muhammad ibn Abdallah al-Tamimi al-Shaffi al-Baghdadi
b. Pathological
c. Independence
d. Plugging in

5. In coordinate geometry, using the common convention that the horizontal axis represents a variable x and the vertical axis represents a variable y, a _____ is a point where the graph of a function or relation intersects with the y-axis of the coordinate system. As such, these points satisfy x=0.

If the curve in question is given as y = f(x), the y-coordinate of the _____ is found by calculating f(0).

a. Y-intercept
b. Point
c. Set
d. Frequency histogram

1. c
2. d
3. b
4. c
5. a

You can take the complete Chapter Practice Test

for Chapter 3. Association: Contingency, Correlation, and Regression
on all key terms, persons, places, and concepts.

Online 99 Cents

http://www.epub27.14.20551.3.cram101.com/

Use www.Cram101.com for all your study needs

including Cram101's online interactive problem solving labs in

chemistry, statistics, mathematics, and more.

CHAPTER OUTLINE: KEY TERMS, PEOPLE, PLACES, CONCEPTS

Response variable

Squared

Correlation

Cell

Cause and effect

Anecdotal evidence

Census

General Social Survey

Sample survey

Statistic

Financial statement

P-value

Random sampling

Sampling design

Sampling frame

Changing-criterion research design

Random number table

Random sample

ANOVA

_____ Variance

_____ Random digit dialing

_____ Telephone interview

_____ Margin of error

_____ Questionnaire

_____ Missing data

_____ Sampling bias

_____ Response bias

_____ Placebo effect

_____ Randomization

_____ Blind method

_____ Clinical trial

_____ Statistical significance

_____ Theorem

_____ Lung

_____ Factorization

_____ Logistic regression

_____ Paradox

_____ Crossover

CHAPTER OUTLINE: KEY TERMS, PEOPLE, PLACES, CONCEPTS

_____   Block

_____   Austin Bradford Hill

_____   Randomized block design

_____   Block design

_____   Bar graph

_____   Categorical variable

_____   Descriptive statistic

_____   Histogram

_____   Pie chart

_____   Box plot

_____   Contingency table

_____   Deviation

_____   Empirical

_____   Five-number summary

_____   Interquartile range

_____   Median

_____   Outlier

_____   Percentile

_____   Position

	Positive number
	Quartile
	Standard deviation
	Regression line

CHAPTER HIGHLIGHTS & NOTES: KEY TERMS, PEOPLE, PLACES, CONCEPTS

Response variable	The terms 'dependent variable' and 'independent variable' are used in similar but subtly different ways in mathematics and statistics as part of the standard terminology in those subjects. They are used to distinguish between two types of quantities being considered, separating them into those available at the start of a process and those being created by it, where the latter (dependent variables) are dependent on the former (independent variables).

The independent variable is typically the variable representing the value being manipulated or changed and the dependent variable is the observed result of the independent variable being manipulated. For example concerning nutrition, the independent variable of daily vitamin C intake (how much vitamin C one consumes) can influence the dependent variable of life expectancy (the average age one attains). Over some period of time, scientists will control the vitamin C intake in a substantial group of people. One part of the group will be given a daily high dose of vitamin C, and the remainder will be given a placebo pill (so that they are unaware of not belonging to the first group) without vitamin C. The scientists will investigate if there is any statistically significant difference in the life span of the people who took the high dose and those who took the placebo (no dose). The goal is to see if the independent variable of high vitamin C dosage has a correlation with the dependent variable of people's life span. The designation independent/dependent is clear in this case, because if a correlation is found, it cannot be that life span has influenced vitamin C intake, but an influence in the other direction is possible. Use in mathematics

In calculus, a function is a map whose action is specified on variables. Take x and y to be two variables. A function f may map x to some expression in x. Assigning $y = f(x)$ gives a relation between x and y. |

If there is some relation specifying y in terms of x, then y is known as a 'dependent variable' (and x is an 'independent variable'). Use in statistics Controlled experiments

In a statistics experiment, the dependent variable is the event studied and expected to change whenever the independent variable is altered.

In the design of experiments, an independent variable's values are controlled or selected by the experimenter to determine its relationship to an observed phenomenon (i.e., the dependent variable). In such an experiment, an attempt is made to find evidence that the values of the independent variable determine the values of the dependent variable. The independent variable can be changed as required, and its values do not represent a problem requiring explanation in an analysis, but are taken simply as given. The dependent variable, on the other hand, usually cannot be directly controlled.

Controlled variables are also important to identify in experiments. They are the variables that are kept constant to prevent their influence on the effect of the independent variable on the dependent. Every experiment has a controlling variable, and it is necessary to not change it, or the results of the experiment won't be valid.

'Extraneous variables' are those that might affect the relationship between the independent and dependent variables. Extraneous variables are usually not theoretically interesting. They are measured in order for the experimenter to compensate for them. For example, an experimenter who wishes to measure the degree to which caffeine intake (the independent variable) influences explicit recall for a word list (the dependent variable) might also measure the participant's age (extraneous variable). She can then use these age data to control for the uninteresting effect of age, clarifying the relationship between caffeine and memory.

In summary:•Independent variables answer the question 'What do I change?'•Dependent variables answer the question 'What do I observe'?'•Controlled variables answer the question 'What do I keep the same?'•Extraneous variables answer the question 'What uninteresting variables might mediate the effect of the IV on the DV?'Alternative terminology in statistics

In statistics, the dependent/independent variable terminology is used more widely than just in relation to controlled experiments. For example the data analysis of two jointly varying quantities may involve treating each in turn as the dependent variable and the other as the independent variable. However, for general usage, the pair response variable and explanatory variable is preferable as quantities treated as 'independent variables' are rarely statistically independent.

Squared	In algebra, the square of a number is that number multiplied by itself. To square a quantity is to multiply it by itself. Its notation is a superscripted '2'; a number x squared is written as x^2.

Chapter 4. Gathering Data

Correlation	In statistics, correlation (often measured as a correlation coefficient, ρ) indicates the strength and direction of a relationship between two random variables. The commonest use refers to a linear relationship. In general statistical usage, correlation or co-relation refers to the departure of two random variables from independence.
Cell	In geometry, a cell is a three-dimensional element that is part of a higher-dimensional object.
	In polytopes
	A cell is a three-dimensional polyhedron element that is part of the boundary of a higher-dimensional polytope, such as a polychoron (4-polytope) or honeycomb (3-space tessellation).
	For example, a cubic honeycomb is made of cubic cells, with 4 cubes on each edge.
Cause and effect	Cause and effect refers to the philosophical concept of causality, in which an action or event will produce a certain response to the action in the form of another event.
Anecdotal evidence	The expression anecdotal evidence has two distinct meanings.
	(1) Evidence in the form of an anecdote or hearsay is called anecdotal if there is doubt about its veracity; the evidence itself is considered untrustworthy.
	(2) Evidence, which may itself be true and verifiable, used to deduce a conclusion which does not follow from it, usually by generalizing from an insufficient amount of evidence.
Census	A census is the procedure of systematically acquiring and recording information about the members of a given population. It is a regularly occurring and official count of a particular population. The term is used mostly in connection with national population and housing censuses; other common censuses include agriculture, business, and traffic censuses.
General Social Survey	The General Social Survey is a sociological survey used to collect data on demographic characteristics and attitudes of residents of the United States. The survey is conducted face-to-face with an in-person interview by the National Opinion Research Center at the University of Chicago, of a randomly-selected sample of adults (18+) who are not institutionalized. The survey was conducted every year from 1972 to 1994 (except in 1979, 1981, and 1992).
Sample survey	Sample survey is a survey of a population made by using only a portion of the population.
Statistic	A statistic is a single measure of some attribute of a sample (e.g. its arithmetic mean value). It is calculated by applying a function (statistical algorithm) to the values of the items comprising the sample which are known together as a set of data.

| Financial statement | A Financial statement is a formal record of the financial activities of a business, person, or other entity. In British English--including United Kingdom company law--a Financial statement is often referred to as an account, although the term Financial statement is also used, particularly by accountants.

For a business enterprise, all the relevant financial information, presented in a structured manner and in a form easy to understand, are called the Financial statements. They typically include four basic Financial statements:

· Balance sheet: also referred to as statement of financial position or condition, reports on a company's assets, liabilities, and Ownership equity at a given point in time. · Income statement: also referred to as Profit and Loss statement , reports on a company's income, expenses, and profits over a period of time. |
| --- | --- |
| P-value | In statistical significance testing, the p-value is the probability of obtaining a test statistic at least as extreme as the one that was actually observed, assuming that the null hypothesis is true. In this context, value a is considered more 'extreme' than b if a is less likely to occur under the null. One often 'rejects the null hypothesis' when the p-value is less than the significance level α, which is often 0.05 or 0.01. When the null hypothesis is rejected, the result is said to be statistically significant. |
| Random sampling | In random sampling every combination of items from the frame, or stratum, has a known probability of occurring, but these probabilities are not necessarily equal. With any form of sampling there is a risk that the sample may not adequately represent the population but with random sampling there is a large body of statistical theory which quantifies the risk and thus enables an appropriate sample size to be chosen. |
| Sampling design | In the theory of finite population sampling, a sampling design specifies for every possible sample its probability of being drawn.

Mathematically, a sampling design is denoted by the function $P(S)$ which gives the probability of drawing a sample S. An example of a sampling design

During Bernoulli sampling, $P(S)$ is given by
$$P(S) = q^{N_{\text{sample}}(S)} \times (1-q)^{(N_{\text{pop}} - N_{\text{sample}}(S))}$$

where for each element q is the probability of being included in the sample and $N_{\text{sample}}(S)$ is the total number of elements in the sample S and N_{pop} is the total number of elements in the population (before sampling commenced). |

Chapter 4. Gathering Data

Sampling frame	In statistics, a sampling frame is the source material or device from which a sample is drawn. It is a list of all those within a population who can be sampled, and may include individuals, households or institutions. Importance of the sampling frame is stressed by Jessen:"Sampling frame types and qualities In the most straightforward case, such as when dealing with a batch of material from a production run, or using a census, it is possible to identify and measure every single item in the population and to include any one of them in our sample; this is known as direct element sampling.
Changing-criterion research design	In a changing-criterion research design a criterion for reinforcement is changed across the experiment to demonstrate the functional relationship between the reinforcement and the behavior. See Mark Dixon's work with a participant using a short video clip to generate a preference for a progressively delayed variable reinforcement over a fixed shorter delay reinforcement in physical therapy.
Random number table	Random number tables have been used in statistics for tasks such as selected random samples. This was much more effective than manually selecting the random samples (with dice, cards, etc).. Nowadays, tables of random numbers have been replaced by computational random number generators.
Random sample	In statistics, a sample is a subject chosen from a population for investigation; a random sample is one chosen by a method involving an unpredictable component. Random sampling can also refer to taking a number of independent observations from the same probability distribution, without involving any real population. The sample usually is not a representative of the population of people from which it was drawn-- this random variation in the results is termed as sampling error.
ANOVA	In statistics, ANOVA is a collection of statistical models, and their associated procedures, in which the observed variance is partitioned into components due to different sources of variation. In its simplest form ANOVA provides a statistical test of whether or not the means of several groups are all equal, and therefore generalizes Student's two-sample t-test to more than two groups. ANOVAs are helpful because they possess a certain advantage over a two-sample t-test. Doing multiple two-sample t-tests would result in a largely increased chance of committing a type I error. For this reason, ANOVAs are useful in comparing three or more means. There are three conceptual classes of such models: · Fixed-effects models assume that the data came from normal populations which may differ only in their means. (Model 1)

	· Random effects models assume that the data describe a hierarchy of different populations whose differences are constrained by the hierarchy. (Model 2) · Mixed-effect models describe the situations where both fixed and random effects are present. (Model 3)
Variance	In probability theory and statistics, the variance is a measure of how far a set of numbers is spread out. It is one of several descriptors of a probability distribution, describing how far the numbers lie from the mean (expected value). In particular, the variance is one of the moments of a distribution.
Random digit dialing	Random digit dialing is a method for selecting people for involvement in telephone statistical surveys by generating telephone numbers at random. Random digit dialing has the advantage that it includes unlisted numbers that would be missed if the numbers were selected from a phone book. In populations where there is a high telephone-ownership rate, it can be a cost efficient way to get complete coverage of a geographic area.
Telephone interview	A telephone interview is an interview conducted over the telephone for employment, journalism, or whatever. Often conducted in the initial interview round of the hiring process, this type of interview allows an employer to screen candidates on the candidates experience, qualifications, and salary expectations pertaining to the position and the company. The telephone interview saves the employers' time and eliminates candidates that are unlikely to meet the company's expectations.
Margin of error	The margin of error is a statistic expressing the amount of random sampling error in a survey's results. The larger the margin of error, the less faith one should have that the poll's reported results are close to the 'true' figures; that is, the figures for the whole population. Margin of error occurs whenever a population is incompletely sampled.
Questionnaire	A questionnaire is a research instrument consisting of a series of questions and other prompts for the purpose of gathering information from respondents. Although they are often designed for statistical analysis of the responses, this is not always the case. The questionnaire was invented by Sir Francis Galton.
Missing data	In statistics, missing data, occur when no data value is stored for the variable in the current observation. Missing data are a common occurrence and can have a significant effect on the conclusions that can be drawn from the data. Missing data can occur because of nonresponse: no information is provided for several items or no information is provided for a whole unit.

Chapter 4. Gathering Data

Sampling bias	In statistics, sampling bias is when a sample is collected in such a way that some members of the intended population are less likely to be included than others. It results in a biased sample, a non-random sample of a population (or non-human factors) in which all individuals, or instances, were not equally likely to have been selected. If this is not accounted for, results can be erroneously attributed to the phenomenon under study rather than to the method of sampling.
Response bias	Response bias is a type of cognitive bias which can affect the results of a statistical survey if respondents answer questions in the way they think the questioner wants them to answer rather than according to their true beliefs. This may occur if the questioner is obviously angling for a particular answer (as in push polling) or if the respondent wishes to please the questioner by answering what appears to be the 'morally right' answer. An example of the latter might be if a woman surveys a man on his attitudes to domestic violence, or someone who obviously cares about the environment asks people how much they value a wilderness area.
Placebo effect	Placebo effect is the term applied by medical science to the therapeutical and healing effects of inert medicines and/or ritualistic or faith healing practices. The placebo effect occurs when a patient takes an inert substance . Experimenters typically use placebos in the context of a clinical trial, in which a 'test group' of patients receives the therapy being tested, and a 'control group' receives the placebo. It can then be determined if results from the 'test' group exceed those due to the placebo effect.
Randomization	Randomization is the process of making something random; this means:•Generating a random permutation of a sequence (such as when shuffling cards).•Selecting a random sample of a population (important in statistical sampling).•Generating random numbers.•Transforming a data stream (such as when using a scrambler in telecommunications).
	Applications
	Randomization is used in statistics and in gambling.
	Statistics
	Randomization is a core principle in statistical theory, whose importance was emphasized by Charles S. Peirce in 'Illustrations of the Logic of Science' (1877-1878) and 'A Theory of Probable Inference' (1883). Randomization-based inference is especially important in experimental design and in survey sampling.
Blind method	The Blind method is a part of the scientific method, used to prevent research outcomes from being influenced by either the placebo effect or the observer bias. To blind a person involved in research (whether a researcher, subject, funder, or other person) is to prevent them from knowing certain information about the process.

Clinical trial	Clinical trials are a set of procedures in medical research and drug development that are conducted to allow safety and efficacy data to be collected for health interventions (e.g., drugs, diagnostics, devices, therapy protocols). These trials can take place only after satisfactory information has been gathered on the quality of the non-clinical safety, and Health Authority/Ethics Committee approval is granted in the country where the trial is taking place.
	Depending on the type of product and the stage of its development, investigators enroll healthy volunteers and/or patients into small pilot studies initially, followed by larger scale studies in patients that often compare the new product with the currently prescribed treatment.
Statistical significance	In statistics, a result is called 'statistically significant' if it is unlikely to have occurred by chance. The phrase test of significance was coined by Ronald Fisher. As used in statistics, significant does not mean important or meaningful, as it does in everyday speech. Research analysts who focus solely on significant results may miss important response patterns which individually may fall under the threshold set for tests of significance. Many researchers urge that tests of significance should always be accompanied by effect-size statistics, which approximate the size and thus the practical importance of the difference.
	The amount of evidence required to accept that an event is unlikely to have arisen by chance is known as the significance level or critical p-value: in traditional Fisherian statistical hypothesis testing, the p-value is the probability of observing data at least as extreme as that observed, given that the null hypothesis is true. If the obtained p-value is small then it can be said either the null hypothesis is false or an unusual event has occurred. P-values do not have any repeat sampling interpretation.
	An alternative (but nevertheless related) statistical hypothesis testing framework is the Neyman-Pearson frequentist school which requires both a null and an alternative hypothesis to be defined and investigates the repeat sampling properties of the procedure, i.e. the probability that a decision to reject the null hypothesis will be made when it is in fact true and should not have been rejected (this is called a 'false positive' or Type I error) and the probability that a decision will be made to accept the null hypothesis when it is in fact false (Type II error). Fisherian p-values are philosophically different from Neyman-Pearson Type I errors. This confusion is unfortunately propagated by many statistics textbooks. Use in practice
	The significance level is usually denoted by the Greek symbol α (lowercase alpha). Popular levels of significance are 10% (0.1), 5% (0.05), 1% (0.01), 0.5% (0.005), and 0.1% (0.001). If a test of significance gives a p-value lower than the significance level α, the null hypothesis is rejected. Such results are informally referred to as 'statistically significant'. For example, if someone argues that 'there's only one chance in a thousand this could have happened by coincidence,' a 0.001 level of statistical significance is being implied.

Chapter 4. Gathering Data

Theorem	In mathematics, a Theorem is a statement which has been proven on the basis of previously established statements, such as other Theorems, and previously accepted statements, such as axioms. The derivation of a Theorem is often interpreted as a proof of the truth of the resulting expression, but different deductive systems can yield other interpretations, depending on the meanings of the derivation rules. Theorems have two components, called the hypotheses and the conclusions.
Lung	Lung (Tibetan: rlung) is a word that means wind or breath. It is a key concept in the Vajrayana traditions of Tibetan Buddhism and as such is part of the symbolic 'twilight language', used to non-conceptually point to a variety of meanings. Lung is a concept that's particularly important to understandings of the subtle body and the Three Vajras (body, speech and mind).
Factorization	In mathematics, Factorization or factoring is the decomposition of an object into a product of other objects, which when multiplied together give the original. For example, the number 15 factors into primes as 3×5, and the polynomial $x^2 - 4$ factors as $(x - 2)(x + 2)$. In all cases, a product of simpler objects is obtained.
Logistic regression	In statistics, logistic regression is a type of regression analysis used for predicting the outcome of a categorical (a variable that can take on a limited number of categories) criterion variable based on one or more predictor variables. Logistic regression can be bi- or multinomial. Binomial or binary logistic regression refers to the instance in which the criterion can take on only two possible outcomes (e.g., 'dead' vs. 'alive', 'success' vs. 'failure', or 'yes' vs. 'no').
Paradox	A Paradox is a true statement or group of statements that leads to a contradiction or a situation which defies intuition. The term is also used for an apparent contradiction that actually expresses a non-dual truth (cf. kÅan, Catuskoti).
Crossover	In genetic algorithms, crossover is a genetic operator used to vary the programming of a chromosome or chromosomes from one generation to the next. It is analogous to reproduction and biological crossover, upon which genetic algorithms are based. Crossover techniques Many crossover techniques exist for organisms which use different data structures to store themselves.
Block	In telecommunications a block is one of:•A group of bits or digits that is transmitted as a unit and that may be encoded for error-control purposes.•A string of records, words, or characters, that for technical or logical purposes are treated as a unit. Blocks (a) are separated by interblock gaps, (b) are delimited by an end-of-block signal, and (c) may contain one or more records.

	A block is usually subjected to some type of block processing, such as multidimensional parity checking, associated with it.
	A block transfer attempt is a coordinated sequence of user and telecommunication system activities undertaken to effect transfer of an individual block from a source user to a destination user.
Austin Bradford Hill	Sir Austin Bradford Hill FRS (8 July 1897 - 18 April 1991), English epidemiologist and statistician, pioneered the randomized clinical trial and, together with Richard Doll, was the first to demonstrate the connection between cigarette smoking and lung cancer. Hill is widely known for pioneering the 'Bradford-Hill' criteria for determining a causal association. Early life Son of Sir Leonard Erskine Hill FRS a distinguished physiologist, Hill was born in London, lived as a child at the family home, Osborne House, Loughton, Essex; and educated at Chigwell School, Essex.
Randomized block design	In the statistical theory of the design of experiments, blocking is the arranging of experimental units in groups (blocks) that are similar to one another. Typically, a blocking factor is a source of variability that is not of primary interest to the experimenter. An example of a blocking factor might be the sex of a patient; by blocking on sex, this source of variability is controlled for, thus leading to greater accuracy. Blocking to 'remove' the effect of nuisance factors For randomized block designs, there is one factor or variable that is of primary interest.
Block design	In combinatorial mathematics, a block design is a set together with a family of subsets (repeated subsets are allowed at times) whose members are chosen to satisfy some set of properties that are deemed useful for a particular application. These applications come from many areas, including experimental design, finite geometry, software testing, cryptography, and algebraic geometry. Many variations have been examined, but the most intensely studied are the balanced incomplete block designs (BIBDs or 2-designs) which historically were related to statistical issues in the design of experiments.
Bar graph	A bar chart or Bar graph is a chart with rectangular bars with lengths proportional to the values that they represent. Bar charts are used for comparing two or more values that were taken over time or on different conditions, usually on small data sets. The bars can be horizontal lines or it can also be used to mass a point of view.
Categorical variable	Categorical Variables

Chapter 4. Gathering Data

	In statistics, a categorical variable is a variable that can take on one of a limited, and usually fixed, number of possible values. Categorical variables are often used to represent categorical data. A categorical variable that can take on exactly two values is termed a binary variable and is typically treated on its own as a special case.
Descriptive statistic	Descriptive statistics are used to describe the main features of a collection of data in quantitative terms. Descriptive statistics are distinguished from inferential statistics (or inductive statistics), in that Descriptive statistics aim to quantitatively summarize a data set, rather than being used to support inferential statements about the population that the data are thought to represent. Even when a data analysis draws its main conclusions using inductive statistical analysis, Descriptive statistics are generally presented along with more formal analyses.
Histogram	In statistics, a histogram is a graphical representation showing a visual impression of the distribution of data. It is an estimate of the probability distribution of a continuous variable and was first introduced by Karl Pearson. A histogram consists of tabular frequencies, shown as adjacent rectangles, erected over discrete intervals (bins), with an area equal to the frequency of the observations in the interval.
Pie chart	A pie chart is a circular chart divided into sectors, illustrating proportion. In a pie chart, the arc length of each sector (and consequently its central angle and area), is proportional to the quantity it represents. When angles are measured with 1 turn as unit then a number of percent is identified with the same number of centiturns.
Box plot	In descriptive statistics, a box plot is a convenient way of graphically depicting groups of numerical data through their five-number summaries: the smallest observation (sample minimum), lower quartile (Q1), median (Q2), upper quartile (Q3), and largest observation (sample maximum). A boxplot may also indicate which observations, if any, might be considered outliers. Boxplots display differences between populations without making any assumptions of the underlying statistical distribution: they are non-parametric.
Contingency table	In statistics, a contingency table is a type of table in a matrix format that displays the (multivariate) frequency distribution of the variables. It is often used to record and analyze the relation between two or more categorical variables.

Deviation	In mathematics and statistics, deviation is a measure of difference between the observed value and the mean. The sign of deviation (positive or negative), reports the direction of that difference (it is larger when the sign is positive, and smaller if it is negative). The magnitude of the value indicates the size of the difference.
Empirical	The word empirical denotes information acquired by means of observation or experimentation. Empirical data are data produced by an observation or experiment. A central concept in modern science and the scientific method is that all evidence must be empirical, or empirically based, that is, dependent on evidence or consequences that are observable by the senses.
Five-number summary	The five-number summary is a descriptive statistic that provides information about a set of observations. It consists of the five most important sample percentiles:•the sample minimum (smallest observation)•the lower quartile or first quartile•the median (middle value)•the upper quartile or third quartile•the sample maximum (largest observation) In order for these statistics to exist the observations must be from a univariate variable that can be measured on an ordinal, interval or ratio scale. Use and representation The five-number summary provides a concise summary of the distribution of the observations.
Interquartile range	In descriptive statistics, the interquartile range also called the midspread or middle fifty, is a measure of statistical dispersion, being equal to the difference between the upper and lower quartiles. $IQR = Q_3 - Q_1$ Unlike (total) range, the interquartile range is a robust statistic, having a breakdown point of 25%, and is thus often preferred to the total range. The IQR is used to build box plots, simple graphical representations of a probability distribution.
Median	In probability theory and statistics, a median is described as the numerical value separating the higher half of a sample, a population, or a probability distribution, from the lower half. The median of a finite list of numbers can be found by arranging all the observations from lowest value to highest value and picking the middle one. If there is an even number of observations, then there is no single middle value; the median is then usually defined to be the mean of the two middle values.
Outlier	In statistics, an outlier is an observation that is numerically distant from the rest of the data. Grubbs defined an outlier as:'

Chapter 4. Gathering Data

An outlying observation, or outlier, is one that appears to deviate markedly from other members of the sample in which it occurs. '

Outliers can occur by chance in any distribution, but they are often indicative either of measurement error or that the population has a heavy-tailed distribution.

Percentile

In statistics and the social sciences, a percentile is the value of a variable below which a certain percent of observations fall. For example, the 20th percentile is the value (or score) below which 20 percent of the observations may be found. The term percentile and the related term percentile rank are often used in the reporting of scores from norm-referenced tests.

Position

In geometry, a position, location, or radius vector, usually denoted $\mathbf{r}$, is a vector which represents the position of a point P in space in relation to an arbitrary reference origin O. It corresponds to the displacement from O to P: $\mathbf{r} = \overrightarrow{OP}$.

The concept typically applies to two- or three-dimensional space, but can be easily generalized to Euclidean spaces with a higher number of dimensions. Applications •In linear algebra, a position vector can be expressed as a linear combination of basis vectors.•The kinematic movement of a point mass can be described by a vector-valued function giving the position $\mathbf{r}(t)$ as a function of the scalar time parameter t. These are used in mechanics and dynamics to keep track of the positions of particles, point masses, or rigid objects.•In differential geometry, position vector fields are used to describe continuous and differentiable space curves, in which case the independent parameter needs not be time, but can be (e.g).

Positive number

Being a positive number is a property of a number which is real, or a member of a subset of real numbers such as rational and integer numbers. A positive number is one that is greater than zero, such as $\sqrt{2}$, 1.414, 1. Zero itself is neither positive nor negative. The non-negative numbers are the numbers that are not negative .

Quartile

In descriptive statistics, a Quartile is any of the three values which divide the sorted data set into four equal parts, so that each part represents one fourth of the sampled population.

In epidemiology, the Quartiles are the four ranges defined by the three values discussed here.

· first Quartile (designated Q_1) = lower Quartile = cuts off lowest 25% of data = 25th percentile · second Quartile (designated Q_2) = median = cuts data set in half = 50th percentile · third Quartile (designated Q_3) = upper Quartile = cuts off highest 25% of data, or lowest 75% = 75th percentile

The difference between the upper and lower Quartiles is called the interQuartile range.

There is no universal agreement on choosing the Quartile values.

The formula for locating the position of the observation at a given percentile, y, with n data points sorted in ascending order is:

$$L_y = (n)\left(\frac{y}{100}\right)$$

· Case 1: If L is a whole number, then the value will be found halfway between positions L and L+1
· Case 2: If L is a decimal, round up to the nearest whole number. (for example, L = 1.2 becomes 2) Example 4. Boxplot (with Quartiles and an interQuartile range) and a probability density function (pdf) of a normal $N(0,1\sigma^2)$ population

One possible rule (employed by the TI-83 calculator boxplot and 1-Var Stats functions) is as follows:

· Use the median to divide the ordered data set into two halves.

Standard deviation	Standard deviation is a widely used measurement of variability or diversity used in statistics and probability theory. It shows how much variation or 'dispersion' there is from the 'average' (mean, or expected/budgeted value). A low standard deviation indicates that the data points tend to be very close to the mean, whereas high standard deviation indicates that the data are spread out over a large range of values.
Regression line	Regression line is a line drawn through a scatterplot of two variables. The line is chosen so that it comes as close to the points as possible.

Chapter 4. Gathering Data

1. In statistics, _____ is when a sample is collected in such a way that some members of the intended population are less likely to be included than others. It results in a biased sample, a non-random sample of a population (or non-human factors) in which all individuals, or instances, were not equally likely to have been selected. If this is not accounted for, results can be erroneously attributed to the phenomenon under study rather than to the method of sampling.

   a. Confirmation bias
   b. Design of experiments
   c. Sampling bias
   d. Lack-of-fit sum of squares

2. In geometry, a _____ is a three-dimensional element that is part of a higher-dimensional object.

   In polytopes

   A _____ is a three-dimensional polyhedron element that is part of the boundary of a higher-dimensional polytope, such as a polychoron (4-polytope) or honeycomb (3-space tessellation).

   For example, a cubic honeycomb is made of cubic _____s, with 4 cubes on each edge.

   a. Convex uniform honeycomb
   b. Disphenoid tetrahedral honeycomb
   c. Gyrated tetrahedral-octahedral honeycomb
   d. Cell

3. _____ refers to the philosophical concept of causality, in which an action or event will produce a certain response to the action in the form of another event.

   a. 1-factor
   b. Cause and effect
   c. Chronology protection conjecture
   d. Dilemma of determinism

4. _____ is a method for selecting people for involvement in telephone statistical surveys by generating telephone numbers at random. _____ has the advantage that it includes unlisted numbers that would be missed if the numbers were selected from a phone book. In populations where there is a high telephone-ownership rate, it can be a cost efficient way to get complete coverage of a geographic area.

   a. Rubric
   b. Self-assessment
   c. Random digit dialing
   d. Stress testing

5. . In statistics, _____ (often measured as a _____ coefficient, ρ) indicates the strength and direction of a relationship between two random variables. The commonest use refers to a linear relationship. In general statistical usage, _____ or co-relation refers to the departure of two random variables from independence.

Chapter 4. Gathering Data

a. Covariance matrix

b. Pearson product-moment correlation coefficient

c. Correlation

d. Sample covariance matrix

1. c
2. d
3. b
4. c
5. c

You can take the complete Chapter Practice Test

for Chapter 4. Gathering Data
on all key terms, persons, places, and concepts.

Online 99 Cents

http://www.epub27.14.20551.4.cram101.com/

Use www.Cram101.com for all your study needs

including Cram101's online interactive problem solving labs in

chemistry, statistics, mathematics, and more.

CHAPTER OUTLINE: KEY TERMS, PEOPLE, PLACES, CONCEPTS

	Probability
	Randomness
	Law of large numbers
	Large number
	Relative frequency
	Bayesian statistics
	Statistic
	Sample space
	Tree diagram
	Probability sample
	Space
	Event
	Treating
	Venn diagram
	Intersection
	Mutually exclusive
	Union
	Conditional probability
	Contingency table

	Binomial
	Sampling without replacement
	Without replacement
	Independence
	Random sampling
	Specificity
	Statistical inference
	Inference

CHAPTER HIGHLIGHTS & NOTES: KEY TERMS, PEOPLE, PLACES, CONCEPTS

Probability	Probability is a way of expressing knowledge or belief that an event will occur or has occurred. The concept has been given an exact mathematical meaning in probability theory, which is used extensively in such areas of study as mathematics, statistics, finance, gambling, science, Artificial intelligence/Machine learning and philosophy to draw conclusions about the likelihood of potential events and the underlying mechanics of complex systems. Interpretations The word probability does not have a consistent direct definition.
Randomness	Randomness has somewhat disparate meanings as used in several different fields. It also has common meanings which may have loose connections with some of those more definite meanings. The Oxford English Dictionary defines 'random' thus:'

Law of large numbers	In probability theory, the law of large numbers is a theorem that describes the result of performing the same experiment a large number of times. According to the law, the average of the results obtained from a large number of trials should be close to the expected value, and will tend to become closer as more trials are performed.
	For example, a single roll of a six-sided die produces one of the numbers 1, 2, 3, 4, 5, or 6, each with equal probability.
Large number	Different cultures used different traditional numeral systems for naming large numbers. The extent of large numbers used varied in each culture.
	One interesting point in using large numbers is the confusion on the term billion and milliard in many countries, and the use of zillion to denote a very large number where precision is not required.
Relative frequency	In a series of observations, or trials, the relative frequency of occurrence of an event E is calculated as the number of times the event E happened over the total number of observations made. The relative frequency density of occurrence of an event is the relative frequency of E divided by the size of the bin used to classify E.
Bayesian statistics	Bayesian statistics is that subset of the entire field of statistics in which the evidence about the true state of the world is expressed in terms of degrees of belief or, more specifically, Bayesian probabilities. Such an interpretation is only one of a number of interpretations of probability and there are many other statistical techniques that are not based on 'degrees of belief'.
	The general set of statistical techniques can be divided into a number of activities, many of which have special 'Bayesian' versions.
Statistic	A statistic is a single measure of some attribute of a sample (e.g. its arithmetic mean value). It is calculated by applying a function (statistical algorithm) to the values of the items comprising the sample which are known together as a set of data.
	More formally, statistical theory defines a statistic as a function of a sample where the function itself is independent of the sample's distribution; that is, the function can be stated before realisation of the data.
Sample space	In probability theory, the sample space, often denoted S, Ω, or U (for 'universe'), of an experiment or random trial is the set of all possible outcomes. For example, if the experiment is tossing a coin, the sample space is the set {head, tail}. For tossing two coins, the sample space is {(head,head), (head,tail), (tail,head), (tail,tail)}.

Chapter 5. Probability in Our Daily Lives

Tree diagram	The term Tree diagram refers to a specific type of diagram that has a unique network topology. It can be seen as a specific type of network diagram, which in turn can be seen as a special kind of cluster diagram.

· In mathematics and statistical methods, a Tree diagram is used to determine the probability of getting specific results where the possibilities are nested. |
| Probability sample | A group of cases selected from a population by a random process is a probability sample. Every member of the population has a known, nonzero probability of being selected. |
| Space | In mathematics, a space is a set with some added structure.

Mathematical spaces often form a hierarchy, i.e., one space may inherit all the characteristics of a parent space. For instance, all inner product spaces are also normed vector spaces, because the inner product induces a norm on the inner product space such that: $\|x\| = \sqrt{\langle x, x \rangle}$.

Modern mathematics treats 'space' quite differently compared to classical mathematics. |
Event	In probability theory, an event is a set of outcomes (a subset of the sample space) to which a probability is assigned. Typically, when the sample space is finite, any subset of the sample space is an event (i.e. all elements of the power set of the sample space are defined as events). However, this approach does not work well in cases where the sample space is uncountably infinite, most notably when the outcome is a real number.
Treating	Treating, in law, is the act of serving food, drink, and other refreshments as a method of influencing people for political gain. In various countries, treating is considered a form of corruption, and is illegal as such.
Venn diagram	Venn diagrams or set diagrams are diagrams that show all possible logical relations between a finite collection of sets (aggregation of things). Venn diagrams were conceived around 1880 by John Venn. They are used to teach elementary set theory, as well as illustrate simple set relationships in probability, logic, statistics, linguistics and computer science .
Intersection	In mathematics, the intersection (denoted as ∩) of two sets A and B is the set that contains all elements of A that also belong to B , but no other elements.

The intersection of A and B is written 'A ∩ B'. Formally: |

Mutually exclusive	In layman's terms, two events are mutually exclusive if they cannot occur at the same time (i.e., they have no common outcomes).
	In logic, two mutually exclusive propositions are propositions that logically cannot both be true. Another term f is 'disjunct.' To say that more than two propositions are mutually exclusive may, depending on context mean that no two of them can both be true, or only that they cannot all be true.
Union	In set theory, the union (denoted as ∪) of a collection of sets is the set of all distinct elements in the collection. The union of a collection of sets $S_1, S_2, S_3, \ldots, S_n$ gives a set $S_1 \cup S_2 \cup S_3 \cup \ldots \cup S_n$.
	The union of two sets A and B is the collection of points which are in A or in B : $$A \cup B = \{x : x \in A \text{ or } x \in B\}$$
	A simple example: $A = \{1,2,3,4\}$ $B = \{5,6,7,8\}$ $$A \cup B = \{1,2,3,4,5,6,7,8\}$$
	Another typical example: $A = \{1,2,3,4,5,6\}$ $B = \{5,6,7,8\}$ $$A \cup B = \{1,2,3,4,5,6,7,8\}$$
	Other more complex operations can be done including the union, if the set is for example defined by a property rather than a finite or assumed infinite enumeration of elements.
Conditional probability	In probability theory, the conditional probability of A given B' is the probability of A if B is known to occur . It is commonly denoted $P(A|B)$, and sometimes $P_B(A)$. (The vertical line should not be mistaken for logical OR).
Contingency table	In statistics, a contingency table is a type of table in a matrix format that displays the (multivariate) frequency distribution of the variables. It is often used to record and analyze the relation between two or more categorical variables.
	The term contingency table was first used by Karl Pearson in 'On the Theory of Contingency and Its Relation to Association and Normal Correlation', part of the Drapers' Company Research Memoirs Biometric Series I published in 1904.
Binomial	In elementary algebra, a Binomial is a polynomial with two terms--the sum of two monomials--often bound by parenthesis or brackets when operated upon. It is the simplest kind of polynomial other than monomials.

· The Binomial $a^2 - b^2$ can be factored as the product of two other Binomials:

$a^2 - b^2 = (a + b)(a - b)$.

This is a special case of the more general formula:

$$a^{n+1} - b^{n+1} = (a - b) \sum_{k=0}^{n} a^k b^{n-k}$$.

· The product of a pair of linear Binomials (ax + b) and (cx + d) is:

$(ax + b)(cx + d) = acx^2 + axd + bcx + bd$.

· A Binomial raised to the nth power, represented as

$(a + b)^n$

can be expanded by means of the Binomial theorem or, equivalently, using Pascal's triangle. Taking a simple example, the perfect square Binomial $(p + q)^2$ can be found by squaring the first term, adding twice the product of the first and second terms and finally adding the square of the second term, to give $p^2 + 2pq + q^2$.

· A simple but interesting application of the cited Binomial formula is the '(m,n)-formula' for generating Pythagorean triples: for m < n, let $a = n^2 - m^2$, $b = 2mn$, $c = n^2 + m^2$, then $a^2 + b^2 = c^2$.

Sampling without replacement	Sampling without replacement means that in each successive trial of an experiment or process, the total number of possible outcomes or the mix of possible outcomes is changed by sampling. The probability of future events is thus changed.
Without replacement	In small populations and often in large ones, simple random sampling is typically done 'without replacement' ('SRSWOR'), i.e., one deliberately avoids choosing any member of the population more than once.

Independence	In probability theory, to say that two events are independent intuitively means that the occurrence of one event makes it neither more nor less probable that the other occurs. For example:•The event of getting a 6 the first time a die is rolled and the event of getting a 6 the second time are independent.•By contrast, the event of getting a 6 the first time a die is rolled and the event that the sum of the numbers seen on the first and second trials is 8 are not independent.•If two cards are drawn with replacement from a deck of cards, the event of drawing a red card on the first trial and that of drawing a red card on the second trial are independent.•By contrast, if two cards are drawn without replacement from a deck of cards, the event of drawing a red card on the first trial and that of drawing a red card on the second trial are again not independent.
	Similarly, two random variables are independent if the conditional probability distribution of either given the observed value of the other is the same as if the other's value had not been observed. The concept of independence extends to dealing with collections of more than two events or random variables.
Random sampling	In random sampling every combination of items from the frame, or stratum, has a known probability of occurring, but these probabilities are not necessarily equal. With any form of sampling there is a risk that the sample may not adequately represent the population but with random sampling there is a large body of statistical theory which quantifies the risk and thus enables an appropriate sample size to be chosen.
Specificity	The specificity is a statistical measure of how well a binary classification test correctly identifies the negative cases, or those cases that do not meet the condition under study. That is, the specificity is the proportion of true negatives of all negative cases in the population. It is a parameter of the test.
Statistical inference	In statistics, statistical inference is the process of drawing conclusions from data subject to random variation, for example, observational errors or sampling variation. More substantially, the terms statistical inference, statistical induction and inferential statistics are used to describe systems of procedures that can be used to draw conclusions from datasets arising from systems affected by random variation, such as observational errors, random sampling, or random experimentation. Initial requirements of such a system of procedures for inference and induction are that the system should produce reasonable answers when applied to well-defined situations and that it should be general enough to be applied across a range of situations.
Inference	Inference is the act of drawing a conclusion by deductive reasoning from given facts. The conclusion drawn is also called an inference. The laws of valid inference are studied in the field of logic.

Chapter 5. Probability in Our Daily Lives

1. In layman's terms, two events are _____ if they cannot occur at the same time (i.e., they have no common outcomes).

   In logic, two _____ propositions are propositions that logically cannot both be true. Another term f is 'disjunct.' To say that more than two propositions are _____ may, depending on context mean that no two of them can both be true, or only that they cannot all be true.

   a. Bayesian probability
   b. Black swan theory
   c. Computer-assisted proof
   d. Mutually exclusive

2. _____s or set diagrams are diagrams that show all possible logical relations between a finite collection of sets (aggregation of things). _____s were conceived around 1880 by John Venn. They are used to teach elementary set theory, as well as illustrate simple set relationships in probability, logic, statistics, linguistics and computer science .

   a. Venn diagram
   b. Weighted Voronoi diagram
   c. Zone diagram
   d. Function composition

3. A group of cases selected from a population by a random process is a _____. Every member of the population has a known, nonzero probability of being selected.

   a. Nomological network
   b. Thematic Apperception Test
   c. Probability sample
   d. Ratio measurement

4. _____, in law, is the act of serving food, drink, and other refreshments as a method of influencing people for political gain. In various countries, _____ is considered a form of corruption, and is illegal as such.

   a. 1-factor
   b. Experiment
   c. Treating
   d. Infinite monkey theorem

5. . _____ is a way of expressing knowledge or belief that an event will occur or has occurred. The concept has been given an exact mathematical meaning in _____ theory, which is used extensively in such areas of study as mathematics, statistics, finance, gambling, science, Artificial intelligence/Machine learning and philosophy to draw conclusions about the likelihood of potential events and the underlying mechanics of complex systems.

   Interpretations

   The word _____ does not have a consistent direct definition.

Chapter 5. Probability in Our Daily Lives

Visit Cram101.com for full Practice Exams

a. 1-factor
b. 4-dimensional Euclidean space
c. Bacterial growth
d. Probability

1. d
2. a
3. c
4. c
5. d

You can take the complete Chapter Practice Test

for Chapter 5. Probability in Our Daily Lives
on all key terms, persons, places, and concepts.

Online 99 Cents

http://www.epub27.14.20551.5.cram101.com/

Use www.Cram101.com for all your study needs

including Cram101's online interactive problem solving labs in

chemistry, statistics, mathematics, and more.

Chapter 6. Probability Distributions

CHAPTER OUTLINE: KEY TERMS, PEOPLE, PLACES, CONCEPTS

Binomial

Binomial distribution

Normal distribution

Probability

Probability distribution

Discrete probability distributions

Random sampling

Random variable

Continuous random variable

ANOVA

Parameter

Expected value

Sample mean

Weighted average

Pareto chart

Categorical variable

Empirical

Cumulative probability

Percentile

	Relative standing
	Standard deviation
	Standard normal distribution
	Deviation
	Binomial coefficient
	Coefficient
	Factorial

CHAPTER HIGHLIGHTS & NOTES: KEY TERMS, PEOPLE, PLACES, CONCEPTS

Binomial	In elementary algebra, a Binomial is a polynomial with two terms--the sum of two monomials--often bound by parenthesis or brackets when operated upon. It is the simplest kind of polynomial other than monomials.
	· The Binomial $a^2 - b^2$ can be factored as the product of two other Binomials:
	$a^2 - b^2 = (a + b)(a - b)$.
	This is a special case of the more general formula:
	$$a^{n+1} - b^{n+1} = (a - b) \sum_{k=0}^{n} a^k b^{n-k}$$
	· The product of a pair of linear Binomials (ax + b) and (cx + d) is:

$(ax + b)(cx + d) = acx^2 + axd + bcx + bd$.

· A Binomial raised to the nth power, represented as

$(a + b)^n$

can be expanded by means of the Binomial theorem or, equivalently, using Pascal's triangle. Taking a simple example, the perfect square Binomial $(p + q)^2$ can be found by squaring the first term, adding twice the product of the first and second terms and finally adding the square of the second term, to give $p^2 + 2pq + q^2$.

· A simple but interesting application of the cited Binomial formula is the '(m,n)-formula' for generating Pythagorean triples: for $m < n$, let $a = n^2 - m^2$, $b = 2mn$, $c = n^2 + m^2$, then $a^2 + b^2 = c^2$.

Binomial distribution	In probability theory and statistics, the binomial distribution is the discrete probability distribution of the number of successes in a sequence of n independent yes/no experiments, each of which yields success with probability p. Such a success/failure experiment is also called a Bernoulli experiment or Bernoulli trial; when n = 1, the binomial distribution is a Bernoulli distribution. The binomial distribution is the basis for the popular binomial test of statistical significance.
Normal distribution	In probability theory, the normal (or Gaussian) distribution is a continuous probability distribution that has a bell-shaped probability density function, known as the Gaussian function or informally the bell curve: $$f(x; \mu, \sigma^2) = \frac{1}{\sigma\sqrt{2\pi}} e^{-\frac{1}{2}\left(\frac{x-\mu}{\sigma}\right)^2}$$ The parameter μ is the mean or expectation (location of the peak) and σ^2 is the variance. σ is known as the standard deviation. The distribution with $\mu = 0$ and $\sigma^2 = 1$ is called the standard normal distribution or the unit normal distribution.
Probability	Probability is a way of expressing knowledge or belief that an event will occur or has occurred. The concept has been given an exact mathematical meaning in probability theory, which is used extensively in such areas of study as mathematics, statistics, finance, gambling, science, Artificial intelligence/Machine learning and philosophy to draw conclusions about the likelihood of potential events and the underlying mechanics of complex systems. Interpretations

Chapter 6. Probability Distributions

Probability distribution	In probability theory, a probability mass, probability density, or probability distribution is a function that describes the probability of a random variable taking certain values.

For a more precise definition one needs to distinguish between discrete and continuous random variables. In the discrete case, one can easily assign a probability to each possible value: when throwing a die, each of the six values 1 to 6 has the probability 1/6. In contrast, when a random variable takes values from a continuum, probabilities are nonzero only if they refer to finite intervals: in quality control one might demand that the probability of a '500 g' package containing between 490 g and 510 g should be no less than 98%. |
| Discrete probability distributions | Discrete probability distributions arise in the mathematical description of probabilistic and statistical problems in which the values that might be observed are restricted to being within a pre-defined list of possible values. This list has either a finite number of members, or at most is countable.

In probability theory, a probability distribution is called discrete if it is characterized by a probability mass function. |
| Random sampling | In random sampling every combination of items from the frame, or stratum, has a known probability of occurring, but these probabilities are not necessarily equal. With any form of sampling there is a risk that the sample may not adequately represent the population but with random sampling there is a large body of statistical theory which quantifies the risk and thus enables an appropriate sample size to be chosen. |
| Random variable | In probability and statistics, a random variable is subject to variations due to chance (i.e. randomness, in a mathematical sense). As opposed to other mathematical variables, a random variable conceptually does not have a single, fixed value (even if unknown); rather, it can take on a set of possible different values, each with an associated probability. The interpretation of a random variable depends on the interpretation of probability:•The objectivist viewpoint: as the outcome of an experiment or event where randomness is involved (e.g. the result of rolling a die, which is a number between 1 and 6, all with equal probability; or the sum of the results of rolling two dice, which is a number between 2 and 12, with some numbers more likely than others).•The subjectivist viewpoint: the formal encoding of one's beliefs about the various potential values of a quantity that is not known with certainty (e.g. a particular person's belief about the net worth of someone like Bill Gates after Internet research on the subject, which might have possible values ranging between about $50 billion and $100 billion, with values near the center more likely).

Random variables can be classified as either discrete (i.e. it may assume any of a specified list of exact values) or as continuous (i.e. |

Chapter 6. Probability Distributions

Continuous random variable	In probability theory, a probability distribution is called continuous if its cumulative distribution function is continuous. This is equivalent to saying that for random variables X with the distribution in question, Pr[X = a] = 0 for all real numbers a, i.e.: the probability that X attains the value a is zero, for any number a. If the distribution of X is continuous then X is called a Continuous random variable.
ANOVA	In statistics, ANOVA is a collection of statistical models, and their associated procedures, in which the observed variance is partitioned into components due to different sources of variation. In its simplest form ANOVA provides a statistical test of whether or not the means of several groups are all equal, and therefore generalizes Student's two-sample t-test to more than two groups. ANOVAs are helpful because they possess a certain advantage over a two-sample t-test. Doing multiple two-sample t-tests would result in a largely increased chance of committing a type I error. For this reason, ANOVAs are useful in comparing three or more means.
	There are three conceptual classes of such models:
	· Fixed-effects models assume that the data came from normal populations which may differ only in their means. (Model 1) · Random effects models assume that the data describe a hierarchy of different populations whose differences are constrained by the hierarchy. (Model 2) · Mixed-effect models describe the situations where both fixed and random effects are present. (Model 3)
Parameter	Parameter can be interpreted in mathematics, logic, linguistics, environmental science and other disciplines.
	In its common meaning, the term is used to identify a characteristic, a feature, a measurable factor that can help in defining a particular system. It is an important element to take into consideration for the evaluation or for the comprehension of an event, a project or any situation.
Expected value	In probability theory, the expected value of a random variable is the weighted average of all possible values that this random variable can take on. The weights used in computing this average correspond to the probabilities in case of a discrete random variable, or densities in case of a continuous random variable. From a rigorous theoretical standpoint, the expected value is the integral of the random variable with respect to its probability measure.
Sample mean	The sample mean or empirical mean and the sample covariance are statistics computed from a collection of data, thought of as being random.
	Given a random sample $X_1, \ldots, X_N$ from an n-dimensional random variable X (i.e., realizations of N independent random variables with the same distribution as X), the sample mean is

$$\bar{\mathbf{x}} = \frac{1}{N} \sum_{k=1}^{N} \mathbf{x}_k.$$

In coordinates, writing the vectors as columns,

$$\mathbf{x}_k = \begin{bmatrix} x_{1k} \\ \vdots \\ x_{nk} \end{bmatrix}, \quad \bar{\mathbf{x}} = \begin{bmatrix} \bar{x}_1 \\ \vdots \\ \bar{x}_n \end{bmatrix},$$

the entries of the sample mean are

$$\bar{x}_i = \frac{1}{N} \sum_{k=1}^{N} x_{ik}, \quad i = 1, \ldots, n.$$

The sample covariance of $\mathbf{x}_1, \ldots, \mathbf{x}_N$ is the n-by-n matrix $\mathbf{Q} = [q_{ij}]$ with the entries given by

$$q_{ij} = \frac{1}{N-1} \sum_{k=1}^{N} (x_{ik} - \bar{x}_i)(x_{jk} - \bar{x}_j)$$

The sample mean and the sample covariance matrix are unbiased estimates of the mean and the covariance matrix of the random variable $\mathbf{X}$. The reason why the sample covariance matrix has $N-1$ in the denominator rather than N is essentially that the population mean E(X) is not known and is replaced by the sample mean $\bar{x}$.

Weighted average	In statistics, given a set of data, X = { x1, x2, ..., xn} and corresponding non-negative weights, W = { w1, w2, ..., wn} the weighted average, is calculated as: Mean = $w_i x_i / w_i$.
Pareto chart	A Pareto chart, is a type of chart that contains both bars and a line graph, where individual values are represented in descending order by bars, and the cumulative total is represented by the line. The left vertical axis is the frequency of occurrence, but it can alternatively represent cost or another important unit of measure.

Categorical variable	Categorical Variables In statistics, a categorical variable is a variable that can take on one of a limited, and usually fixed, number of possible values. Categorical variables are often used to represent categorical data. A categorical variable that can take on exactly two values is termed a binary variable and is typically treated on its own as a special case.
Empirical	The word empirical denotes information acquired by means of observation or experimentation. Empirical data are data produced by an observation or experiment. A central concept in modern science and the scientific method is that all evidence must be empirical, or empirically based, that is, dependent on evidence or consequences that are observable by the senses.
Cumulative probability	Cumulative probability is the probability of reaching a certain value or interval plus the probability of all the intervals below the current interval.
Percentile	In statistics and the social sciences, a percentile is the value of a variable below which a certain percent of observations fall. For example, the 20th percentile is the value (or score) below which 20 percent of the observations may be found. The term percentile and the related term percentile rank are often used in the reporting of scores from norm-referenced tests.
Relative standing	Relative standing is a measurement of numbers which indicate where a particular values lies in relation to the rest of the values in a set of data or population.
Standard deviation	Standard deviation is a widely used measurement of variability or diversity used in statistics and probability theory. It shows how much variation or 'dispersion' there is from the 'average' (mean, or expected/budgeted value). A low standard deviation indicates that the data points tend to be very close to the mean, whereas high standard deviation indicates that the data are spread out over a large range of values.
Standard normal distribution	The standard normal distribution is the normal distribution with a mean of zero and a standard deviation of one. It is often called the bell curve because the graph of its probability density resembles a bell.
Deviation	In mathematics and statistics, deviation is a measure of difference between the observed value and the mean. The sign of deviation (positive or negative), reports the direction of that difference (it is larger when the sign is positive, and smaller if it is negative).

Chapter 6. Probability Distributions

Binomial coefficient	In mathematics, binomial coefficients are a family of positive integers that occur as coefficients in the binomial theorem. They are indexed by two nonnegative integers; the binomial coefficient indexed by n and k is usually written $\binom{n}{k}$, and it is the coefficient of the x^k term in the polynomial expansion of the binomial power $(1 + x)^n$. Arranging binomial coefficients into rows for successive values of n, and in which k ranges from 0 to n, gives a triangular array called Pascal's triangle.
Coefficient	In mathematics, a Coefficient is a multiplicative factor in some term of an expression (or of a series); it is usually a number, but in any case does not involve any variables of the expression. For instance in $7x^2 - 3xy + 1.5 + y$ the first three terms respectively have Coefficients 7, −3, and 1.5 (in the third term there are no variables, so the Coefficient is the term itself; it is called the constant term or constant Coefficient of this expression). The final term does not have any explicitly written Coefficient, but is usually considered to have Coefficient 1, since multiplying by that factor would not change the term.
Factorial	In mathematics, the factorial of a positive integer n, denoted by n!, is the product of all positive integers less than or equal to n. For example, $5! = 5 \times 4 \times 3 \times 2 \times 1 = 120$ 0! is a special case that is explicitly defined to be 1. The factorial operation is encountered in many different areas of mathematics, notably in combinatorics, algebra and mathematical analysis.

Chapter 6. Probability Distributions

1. In mathematics, a _____ is a multiplicative factor in some term of an expression (or of a series); it is usually a number, but in any case does not involve any variables of the expression. For instance in

    $7x^2 - 3xy + 1.5 + y$

    the first three terms respectively have _____s 7, −3, and 1.5 (in the third term there are no variables, so the _____ is the term itself; it is called the constant term or constant _____ of this expression). The final term does not have any explicitly written _____, but is usually considered to have _____ 1, since multiplying by that factor would not change the term.

    a. Determinant
    b. Coefficient
    c. Delta invariant
    d. Block Lanczos algorithm for nullspace of a matrix over a finite field

2. . In elementary algebra, a _____ is a polynomial with two terms--the sum of two monomials--often bound by parenthesis or brackets when operated upon. It is the simplest kind of polynomial other than monomials.

    · The _____ $a^2 - b^2$ can be factored as the product of two other _____s:

    $a^2 - b^2 = (a + b)(a - b)$.

    $$a^{n+1} - b^{n+1} = (a - b) \sum_{k=0}^{n} a^k\, b^{n-k}$$

    This is a special case of the more general formula: .

    · The product of a pair of linear _____s (ax + b) and (cx + d) Is:

    $(ax + b)(cx + d) = acx^2 + axd + bcx + bd$.

    · A _____ raised to the nth power, represented as

    $(a + b)^n$

    can be expanded by means of the _____ theorem or, equivalently, using Pascal's triangle. Taking a simple example, the perfect square _____ $(p + q)^2$ can be found by squaring the first term, adding twice the product of the first and second terms and finally adding the square of the second term, to give $p^2 + 2pq + q^2$.

    · A simple but interesting application of the cited _____ formula is the '(m,n)-formula' for generating Pythagorean triples: for m < n, let $a = n^2 - m^2$, $b = 2mn$, $c = n^2 + m^2$, then $a^2 + b^2 = c^2$.

Chapter 6. Probability Distributions

    a. 1-factor
    b. 4-dimensional Euclidean space
    c. Binomial
    d. Balanced design

3. In statistics and the social sciences, a _____ is the value of a variable below which a certain percent of observations fall. For example, the 20th _____ is the value (or score) below which 20 percent of the observations may be found. The term _____ and the related term _____ rank are often used in the reporting of scores from norm-referenced tests.

    a. Polychoric correlation
    b. Quantile
    c. Percentile
    d. Summary statistic

4. In probability theory and statistics, the _____ is the discrete probability distribution of the number of successes in a sequence of n independent yes/no experiments, each of which yields success with probability p. Such a success/failure experiment is also called a Bernoulli experiment or Bernoulli trial; when n = 1, the _____ is a Bernoulli distribution. The _____ is the basis for the popular binomial test of statistical significance.

    a. 1-factor
    b. 4-dimensional Euclidean space
    c. Binomial distribution
    d. Balanced design

5. In probability theory, the normal (or Gaussian) distribution is a continuous probability distribution that has a bell-shaped probability density function, known as the Gaussian function or informally the bell curve:

$$f(x; \mu, \sigma^2) = \frac{1}{\sigma\sqrt{2\pi}} e^{-\frac{1}{2}\left(\frac{x-\mu}{\sigma}\right)^2}$$

The parameter μ is the mean or expectation (location of the peak) and σ^2 is the variance. σ is known as the standard deviation. The distribution with μ = 0 and σ^2 = 1 is called the standard _____ or the unit _____.

    a. Poisson distribution
    b. Binomial distribution
    c. Normal distribution
    d. 4-dimensional Euclidean space

1. b
2. c
3. c
4. c
5. c

You can take the complete Chapter Practice Test

for Chapter 6. Probability Distributions
on all key terms, persons, places, and concepts.

Online 99 Cents

http://www.epub27.14.20551.6.cram101.com/

Use www.Cram101.com for all your study needs

including Cram101's online interactive problem solving labs in

chemistry, statistics, mathematics, and more.

Chapter 7. Sampling Distributions

CHAPTER OUTLINE: KEY TERMS, PEOPLE, PLACES, CONCEPTS

_____ | Parameter

_____ | Statistic

_____ | Random sampling

_____ | Sample proportion

_____ | Sampling distribution

_____ | Variability sampling

_____ | ANOVA

_____ | Variance

_____ | Standard deviation

_____ | Deviation

_____ | Normal distribution

_____ | Population mean

_____ | Sample mean

_____ | Central limit theorem

_____ | Theorem

_____ | Law of large numbers

_____ | Large number

_____ | Inference

_____ | Binomial

Binomial distribution

Election

Event

Probability

Continuous random variable

Independence

Probability distribution

Conditional probability

Random variable

Statistical inference

Standard normal distribution

| Parameter | Parameter can be interpreted in mathematics, logic, linguistics, environmental science and other disciplines.

In its common meaning, the term is used to identify a characteristic, a feature, a measurable factor that can help in defining a particular system. It is an important element to take into consideration for the evaluation or for the comprehension of an event, a project or any situation. |
| --- | --- |
| Statistic | A statistic is a single measure of some attribute of a sample (e.g. its arithmetic mean value). It is calculated by applying a function (statistical algorithm) to the values of the items comprising the sample which are known together as a set of data.

More formally, statistical theory defines a statistic as a function of a sample where the function itself is independent of the sample's distribution; that is, the function can be stated before realisation of the data. |
Random sampling	In random sampling every combination of items from the frame, or stratum, has a known probability of occurring, but these probabilities are not necessarily equal. With any form of sampling there is a risk that the sample may not adequately represent the population but with random sampling there is a large body of statistical theory which quantifies the risk and thus enables an appropriate sample size to be chosen.
Sample proportion	Sample proportion is the fraction of samples which were successes. The proportion of successes in the sample is also a random variable.
Sampling distribution	In statistics, a sampling distribution is the probability distribution of a given statistic based on a random sample. Sampling distributions are important in statistics because they provide a major simplification on the route to statistical inference. More specifically, they allow analytical considerations to be based on the sampling distribution of a statistic, rather than on the joint probability distribution of all the individual sample values.
Variability sampling	Variability sampling refers to the different values which a given function of the data takes when it is computed for two or more samples dawn from the same population.
ANOVA	In statistics, ANOVA is a collection of statistical models, and their associated procedures, in which the observed variance is partitioned into components due to different sources of variation. In its simplest form ANOVA provides a statistical test of whether or not the means of several groups are all equal, and therefore generalizes Student's two-sample t-test to more than two groups. ANOVAs are helpful because they possess a certain advantage over a two-sample t-test. Doing multiple two-sample t-tests would result in a largely increased chance of committing a type I error. For this reason, ANOVAs are useful in comparing three or more means.

Chapter 7. Sampling Distributions

	There are three conceptual classes of such models: · Fixed-effects models assume that the data came from normal populations which may differ only in their means. (Model 1) · Random effects models assume that the data describe a hierarchy of different populations whose differences are constrained by the hierarchy. (Model 2) · Mixed-effect models describe the situations where both fixed and random effects are present. (Model 3)
Variance	In probability theory and statistics, the variance is a measure of how far a set of numbers is spread out. It is one of several descriptors of a probability distribution, describing how far the numbers lie from the mean (expected value). In particular, the variance is one of the moments of a distribution.
Standard deviation	Standard deviation is a widely used measurement of variability or diversity used in statistics and probability theory. It shows how much variation or 'dispersion' there is from the 'average' (mean, or expected/budgeted value). A low standard deviation indicates that the data points tend to be very close to the mean, whereas high standard deviation indicates that the data are spread out over a large range of values.
Deviation	In mathematics and statistics, deviation is a measure of difference between the observed value and the mean. The sign of deviation (positive or negative), reports the direction of that difference (it is larger when the sign is positive, and smaller if it is negative). The magnitude of the value indicates the size of the difference.
Normal distribution	In probability theory, the normal (or Gaussian) distribution is a continuous probability distribution that has a bell-shaped probability density function, known as the Gaussian function or informally the bell curve: $$f(x; \mu, \sigma^2) = \frac{1}{\sigma\sqrt{2\pi}} e^{-\frac{1}{2}\left(\frac{x-\mu}{\sigma}\right)^2}$$ The parameter μ is the mean or expectation (location of the peak) and $\sigma^{?2}$ is the variance. σ is known as the standard deviation. The distribution with $\mu = 0$ and $\sigma^{?2} = 1$ is called the standard normal distribution or the unit normal distribution.
Population mean	The mean of a population has an expected value of μ, known as the population mean. The sample mean makes a good estimator of the population mean, as its expected value is the same as the population mean. The sample mean of a population is a random variable, not a constant, and consequently it will have its own distribution. For a random sample of n observations from a normally distributed population, the sample mean distribution is

$$\bar{x} \sim N \left\{ \mu, \frac{\sigma^2}{n} \right\}.$$

Often, since the population variance is an unknown parameter, it is estimated by the mean sum of squares, which changes the distribution of the sample mean from a normal distribution to a Student's t distribution with n − 1 degrees of freedom.

Sample mean

The sample mean or empirical mean and the sample covariance are statistics computed from a collection of data, thought of as being random.

Given a random sample $\mathbf{x}_1, \ldots, \mathbf{x}_N$ from an n-dimensional random variable $\mathbf{X}$ (i.e., realizations of N independent random variables with the same distribution as $\mathbf{X}$), the sample mean is

$$\bar{\mathbf{x}} = \frac{1}{N} \sum_{k=1}^{N} \mathbf{x}_k.$$

In coordinates, writing the vectors as columns,

$$\mathbf{x}_k = \left[\begin{array}{c} x_{1k} \\ \vdots \\ x_{nk} \end{array} \right], \quad \bar{\mathbf{x}} = \left[\begin{array}{c} \bar{x}_1 \\ \vdots \\ \bar{x}_n \end{array} \right],$$

the entries of the sample mean are

$$\bar{x}_i = \frac{1}{N} \sum_{k=1}^{N} x_{ik}, \quad i = 1, \ldots, n.$$

The sample covariance of $\mathbf{x}_1, \ldots, \mathbf{x}_N$ is the n-by-n matrix $\mathbf{Q} = [q_{ij}]$ with the entries given by

$$q_{ij} = \frac{1}{N-1} \sum_{k=1}^{N} \left(x_{ik} - \bar{x}_i \right) \left(x_{jk} - \bar{x}_j \right)$$

Chapter 7. Sampling Distributions

	The sample mean and the sample covariance matrix are unbiased estimates of the mean and the covariance matrix of the random variable X. The reason why the sample covariance matrix has $N-1$ in the denominator rather than N is essentially that the population mean E(X) is not known and is replaced by the sample mean $\bar{x}$.
Central limit theorem	In probability theory, the central limit theorem states that, given certain conditions, the mean of a sufficiently large number of independent random variables, each with finite mean and variance, will be approximately normally distributed. The central limit theorem has a number of variants. In its common form, the random variables must be identically distributed.
Theorem	In mathematics, a Theorem is a statement which has been proven on the basis of previously established statements, such as other Theorems, and previously accepted statements, such as axioms. The derivation of a Theorem is often interpreted as a proof of the truth of the resulting expression, but different deductive systems can yield other interpretations, depending on the meanings of the derivation rules. Theorems have two components, called the hypotheses and the conclusions.
Law of large numbers	In probability theory, the law of large numbers is a theorem that describes the result of performing the same experiment a large number of times. According to the law, the average of the results obtained from a large number of trials should be close to the expected value, and will tend to become closer as more trials are performed. For example, a single roll of a six-sided die produces one of the numbers 1, 2, 3, 4, 5, or 6, each with equal probability.
Large number	Different cultures used different traditional numeral systems for naming large numbers. The extent of large numbers used varied in each culture. One interesting point in using large numbers is the confusion on the term billion and milliard in many countries, and the use of zillion to denote a very large number where precision is not required.
Inference	Inference is the act of drawing a conclusion by deductive reasoning from given facts. The conclusion drawn is also called an inference. The laws of valid inference are studied in the field of logic.
Binomial	In elementary algebra, a Binomial is a polynomial with two terms--the sum of two monomials--often bound by parenthesis or brackets when operated upon. It is the simplest kind of polynomial other than monomials.

· The Binomial $a^2 - b^2$ can be factored as the product of two other Binomials:

$a^2 - b^2 = (a + b)(a - b)$.

This is a special case of the more general formula:

$$a^{n+1} - b^{n+1} = (a - b) \sum_{k=0}^{n} a^k b^{n-k}$$.

· The product of a pair of linear Binomials $(ax + b)$ and $(cx + d)$ is:

$(ax + b)(cx + d) = acx^2 + axd + bcx + bd$.

· A Binomial raised to the nth power, represented as

$(a + b)^n$

can be expanded by means of the Binomial theorem or, equivalently, using Pascal's triangle. Taking a simple example, the perfect square Binomial $(p + q)^2$ can be found by squaring the first term, adding twice the product of the first and second terms and finally adding the square of the second term, to give $p^2 + 2pq + q^2$.

· A simple but interesting application of the cited Binomial formula is the '(m,n)-formula' for generating Pythagorean triples: for m < n, let $a = n^2 - m^2$, $b = 2mn$, $c = n^2 + m^2$, then $a^2 + b^2 = c^2$.

Binomial distribution	In probability theory and statistics, the binomial distribution is the discrete probability distribution of the number of successes in a sequence of n independent yes/no experiments, each of which yields success with probability p. Such a success/failure experiment is also called a Bernoulli experiment or Bernoulli trial; when n = 1, the binomial distribution is a Bernoulli distribution. The binomial distribution is the basis for the popular binomial test of statistical significance.
Election	An election is a formal decision-making process by which a population chooses an individual to hold public office. Elections have been the usual mechanism by which modern representative democracy operates since the 17th century.

Chapter 7. Sampling Distributions

Event	In probability theory, an event is a set of outcomes (a subset of the sample space) to which a probability is assigned. Typically, when the sample space is finite, any subset of the sample space is an event (i.e. all elements of the power set of the sample space are defined as events). However, this approach does not work well in cases where the sample space is uncountably infinite, most notably when the outcome is a real number.
Probability	Probability is a way of expressing knowledge or belief that an event will occur or has occurred. The concept has been given an exact mathematical meaning in probability theory, which is used extensively in such areas of study as mathematics, statistics, finance, gambling, science, Artificial intelligence/Machine learning and philosophy to draw conclusions about the likelihood of potential events and the underlying mechanics of complex systems. Interpretations The word probability does not have a consistent direct definition.
Continuous random variable	In probability theory, a probability distribution is called continuous if its cumulative distribution function is continuous. This is equivalent to saying that for random variables X with the distribution in question, $Pr[X = a] = 0$ for all real numbers a, i.e.: the probability that X attains the value a is zero, for any number a. If the distribution of X is continuous then X is called a Continuous random variable.
Independence	In probability theory, to say that two events are independent intuitively means that the occurrence of one event makes it neither more nor less probable that the other occurs. For example:•The event of getting a 6 the first time a die is rolled and the event of getting a 6 the second time are independent.•By contrast, the event of getting a 6 the first time a die is rolled and the event that the sum of the numbers seen on the first and second trials is 8 are not independent.•If two cards are drawn with replacement from a deck of cards, the event of drawing a red card on the first trial and that of drawing a red card on the second trial are independent.•By contrast, if two cards are drawn without replacement from a deck of cards, the event of drawing a red card on the first trial and that of drawing a red card on the second trial are again not independent. Similarly, two random variables are independent if the conditional probability distribution of either given the observed value of the other is the same as if the other's value had not been observed. The concept of independence extends to dealing with collections of more than two events or random variables.
Probability distribution	In probability theory, a probability mass, probability density, or probability distribution is a function that describes the probability of a random variable taking certain values.

For a more precise definition one needs to distinguish between discrete and continuous random variables. In the discrete case, one can easily assign a probability to each possible value: when throwing a die, each of the six values 1 to 6 has the probability 1/6. In contrast, when a random variable takes values from a continuum, probabilities are nonzero only if they refer to finite intervals: in quality control one might demand that the probability of a '500 g' package containing between 490 g and 510 g should be no less than 98%.

Conditional probability	In probability theory, the conditional probability of A given B ' is the probability of A if B is known to occur . It is commonly denoted $P(A|B)$, and sometimes $P_B(A)$. (The vertical line should not be mistaken for logical OR).
Random variable	In probability and statistics, a random variable is subject to variations due to chance (i.e. randomness, in a mathematical sense). As opposed to other mathematical variables, a random variable conceptually does not have a single, fixed value (even if unknown); rather, it can take on a set of possible different values, each with an associated probability. The interpretation of a random variable depends on the interpretation of probability:•The objectivist viewpoint: as the outcome of an experiment or event where randomness is involved (e.g. the result of rolling a die, which is a number between 1 and 6, all with equal probability; or the sum of the results of rolling two dice, which is a number between 2 and 12, with some numbers more likely than others).•The subjectivist viewpoint: the formal encoding of one's beliefs about the various potential values of a quantity that is not known with certainty (e.g. a particular person's belief about the net worth of someone like Bill Gates after Internet research on the subject, which might have possible values ranging between about $50 billion and $100 billion, with values near the center more likely). Random variables can be classified as either discrete (i.e. it may assume any of a specified list of exact values) or as continuous (i.e. it may assume any numerical value in an interval or collection of intervals).
Statistical inference	In statistics, statistical inference is the process of drawing conclusions from data subject to random variation, for example, observational errors or sampling variation. More substantially, the terms statistical inference, statistical induction and inferential statistics are used to describe systems of procedures that can be used to draw conclusions from datasets arising from systems affected by random variation, such as observational errors, random sampling, or random experimentation. Initial requirements of such a system of procedures for inference and induction are that the system should produce reasonable answers when applied to well-defined situations and that it should be general enough to be applied across a range of situations.
Standard normal distribution	The standard normal distribution is the normal distribution with a mean of zero and a standard deviation of one.

Chapter 7. Sampling Distributions

1. In mathematics and statistics, deviation is a measure of difference between the observed value and the mean. The sign of _____(positive or negative), reports the direction of that difference (it is larger when the sign is positive, and smaller if it is negative). The magnitude of the value indicates the size of the difference.

   a. Deviation
   b. Floor effect
   c. Grand mean
   d. Homogeneity

2. An _____ is a formal decision-making process by which a population chooses an individual to hold public office. _____s have been the usual mechanism by which modern representative democracy operates since the 17th century. _____s may fill offices in the legislature, sometimes in the executive and judiciary, and for regional and local government.

   a. Ellsberg paradox
   b. IDF model
   c. Unanimity
   d. Election

3. . In elementary algebra, a _____ is a polynomial with two terms--the sum of two monomials--often bound by parenthesis or brackets when operated upon. It is the simplest kind of polynomial other than monomials.

   · The _____ $a^2 - b^2$ can be factored as the product of two other _____s:

   $a^2 - b^2 = (a + b)(a - b)$.

   This is a special case of the more general formula:

   $$a^{n+1} - b^{n+1} = (a - b) \sum_{k=0}^{n} a^k b^{n-k}$$.

   · The product of a pair of linear _____s (ax + b) and (cx + d) is:

   $(ax + b)(cx + d) = acx^2 + axd + bcx + bd$.

   · A _____ raised to the nth power, represented as

   $(a + b)^n$

   can be expanded by means of the _____ theorem or, equivalently, using Pascal's triangle.

Taking a simple example, the perfect square _____ $(p + q)^2$ can be found by squaring the first term, adding twice the product of the first and second terms and finally adding the square of the second term, to give $p^2 + 2pq + q^2$.

· A simple but interesting application of the cited _____ formula is the '(m,n)-formula' for generating Pythagorean triples: for m < n, let $a = n^2 - m^2$, $b = 2mn$, $c = n^2 + m^2$, then $a^2 + b^2 = c^2$.

a. Binomial
b. Analytic reasoning
c. Automated reasoning
d. Adaptive reasoning

4. In probability theory, a probability distribution is called continuous if its cumulative distribution function is continuous. This is equivalent to saying that for random variables X with the distribution in question, $\Pr[X = a] = 0$ for all real numbers a, i.e.: the probability that X attains the value a is zero, for any number a. If the distribution of X is continuous then X is called a _____.

a. sample space
b. density function
c. Continuous random variable
d. Coefficient of variation

5. In _____ every combination of items from the frame, or stratum, has a known probability of occurring, but these probabilities are not necessarily equal. With any form of sampling there is a risk that the sample may not adequately represent the population but with _____ there is a large body of statistical theory which quantifies the risk and thus enables an appropriate sample size to be chosen.

a. 1-factor
b. Jet
c. Lebrun manifold
d. Random sampling

1. a
2. d
3. a
4. c
5. d

You can take the complete Chapter Practice Test

for Chapter 7. Sampling Distributions
on all key terms, persons, places, and concepts.

Online 99 Cents

http://www.epub27.14.20551.7.cram101.com/

Use www.Cram101.com for all your study needs

including Cram101's online interactive problem solving labs in

chemistry, statistics, mathematics, and more.

Chapter 8. Statistical Inference: Confidence Intervals

CHAPTER OUTLINE: KEY TERMS, PEOPLE, PLACES, CONCEPTS

	Parameter
	Squared
	Statistical inference
	Statistic
	Correlation
	Inference
	Unbiased estimator
	Estimator
	Confidence interval
	Margin of error
	Standard error
	Probability
	Higher
	Random sampling
	Binomial
	Performance
	Population mean
	Standard deviation
	Deviation

Student t distribution

Standard normal distribution

Normal distribution

Financial statement

Outlier

Relative standing

Gosset

P-value

Estimation

Exact test

ANOVA

Factorization

Ronald Aylmer Fisher

Chapter 8. Statistical Inference: Confidence Intervals

Parameter	Parameter can be interpreted in mathematics, logic, linguistics, environmental science and other disciplines. In its common meaning, the term is used to identify a characteristic, a feature, a measurable factor that can help in defining a particular system. It is an important element to take into consideration for the evaluation or for the comprehension of an event, a project or any situation.
Squared	In algebra, the square of a number is that number multiplied by itself. To square a quantity is to multiply it by itself. Its notation is a superscripted '2'; a number x squared is written as x^2.
Statistical inference	In statistics, statistical inference is the process of drawing conclusions from data subject to random variation, for example, observational errors or sampling variation. More substantially, the terms statistical inference, statistical induction and inferential statistics are used to describe systems of procedures that can be used to draw conclusions from datasets arising from systems affected by random variation, such as observational errors, random sampling, or random experimentation. Initial requirements of such a system of procedures for inference and induction are that the system should produce reasonable answers when applied to well-defined situations and that it should be general enough to be applied across a range of situations.
Statistic	A statistic is a single measure of some attribute of a sample (e.g. its arithmetic mean value). It is calculated by applying a function (statistical algorithm) to the values of the items comprising the sample which are known together as a set of data. More formally, statistical theory defines a statistic as a function of a sample where the function itself is independent of the sample's distribution; that is, the function can be stated before realisation of the data.
Correlation	In statistics, correlation (often measured as a correlation coefficient, ρ) indicates the strength and direction of a relationship between two random variables. The commonest use refers to a linear relationship. In general statistical usage, correlation or co-relation refers to the departure of two random variables from independence.
Inference	Inference is the act of drawing a conclusion by deductive reasoning from given facts. The conclusion drawn is also called an inference. The laws of valid inference are studied in the field of logic.
Unbiased estimator	Unbiased estimator is an estimator that has its expected value the parametric value. An estimator or decision rule having nonzero bias is said to be biased.
Estimator	In statistics, an estimator is a rule for calculating an estimate of a given quantity based on observed data: thus the rule and its result (the estimate) are distinguished.

	There are point and interval estimators. The point estimators yield single-valued results, although this includes the possibility of single vector-valued results and results that can be expressed as a single function.
Confidence interval	In statistics, a confidence interval is a kind of interval estimate of a population parameter and is used to indicate the reliability of an estimate. It is an observed interval (i.e. it is calculated from the observations), in principle different from sample to sample, that frequently includes the parameter of interest, if the experiment is repeated. How frequently the observed interval contains the parameter is determined by the confidence level or confidence coefficient.
Margin of error	The margin of error is a statistic expressing the amount of random sampling error in a survey's results. The larger the margin of error, the less faith one should have that the poll's reported results are close to the 'true' figures; that is, the figures for the whole population. Margin of error occurs whenever a population is incompletely sampled.
Standard error	The standard error is the standard deviation of the sampling distribution of a statistic. The term may also be used to refer to an estimate of that standard deviation, derived from a particular sample used to compute the estimate. For example, the sample mean is the usual estimator of a population mean.
Probability	Probability is a way of expressing knowledge or belief that an event will occur or has occurred. The concept has been given an exact mathematical meaning in probability theory, which is used extensively in such areas of study as mathematics, statistics, finance, gambling, science, Artificial intelligence/Machine learning and philosophy to draw conclusions about the likelihood of potential events and the underlying mechanics of complex systems. Interpretations The word probability does not have a consistent direct definition.
Higher	In Scotland the Higher (Scottish Gaelic: An Àrd Ìre) is one of the national school-leaving certificate exams and university entrance qualifications of the Scottish Qualifications Certificate (SQC) offered by the Scottish Qualifications Authority. It superseded the old Higher Grade on the Scottish Certificate of Education (SCE). Both are normally referred to simply as 'Highers'.
Random sampling	In random sampling every combination of items from the frame, or stratum, has a known probability of occurring, but these probabilities are not necessarily equal.

Chapter 8. Statistical Inference: Confidence Intervals

Binomial	In elementary algebra, a Binomial is a polynomial with two terms--the sum of two monomials--often bound by parenthesis or brackets when operated upon. It is the simplest kind of polynomial other than monomials.

· The Binomial $a^2 - b^2$ can be factored as the product of two other Binomials:

$a^2 - b^2 = (a + b)(a - b)$.

This is a special case of the more general formula:

$$a^{n+1} - b^{n+1} = (a - b) \sum_{k=0}^{n} a^k b^{n-k}$$

· The product of a pair of linear Binomials (ax + b) and (cx + d) is:

$(ax + b)(cx + d) = acx^2 + axd + bcx + bd$.

· A Binomial raised to the nth power, represented as

$(a + b)^n$

can be expanded by means of the Binomial theorem or, equivalently, using Pascal's triangle. Taking a simple example, the perfect square Binomial $(p + q)^2$ can be found by squaring the first term, adding twice the product of the first and second terms and finally adding the square of the second term, to give $p^2 + 2pq + q^2$.

· A simple but interesting application of the cited Binomial formula is the '(m,n)-formula' for generating Pythagorean triples: for m < n, let $a = n^2 - m^2$, $b = 2mn$, $c = n^2 + m^2$, then $a^2 + b^2 = c^2$.

Performance	A performance, in performing arts, generally comprises an event in which one group of people (the performer or performers) behave in a particular way for another group of people (the audience). Sometimes the dividing line between performer and the audience may become blurred, as in the example of 'participatory theatre' where audience members might get involved in the production.

Population mean	The mean of a population has an expected value of µ, known as the population mean. The sample mean makes a good estimator of the population mean, as its expected value is the same as the population mean. The sample mean of a population is a random variable, not a constant, and consequently it will have its own distribution. For a random sample of n observations from a normally distributed population, the sample mean distribution is $$\bar{x} \sim N\left\{\mu, \frac{\sigma^2}{n}\right\}.$$ Often, since the population variance is an unknown parameter, it is estimated by the mean sum of squares, which changes the distribution of the sample mean from a normal distribution to a Student's t distribution with n − 1 degrees of freedom.
Standard deviation	Standard deviation is a widely used measurement of variability or diversity used in statistics and probability theory. It shows how much variation or 'dispersion' there is from the 'average' (mean, or expected/budgeted value). A low standard deviation indicates that the data points tend to be very close to the mean, whereas high standard deviation indicates that the data are spread out over a large range of values.
Deviation	In mathematics and statistics, deviation is a measure of difference between the observed value and the mean. The sign of deviation (positive or negative), reports the direction of that difference (it is larger when the sign is positive, and smaller if it is negative). The magnitude of the value indicates the size of the difference.
Student t distribution	The Student t distribution is a probability distribution that arises in the problem of estimating the mean of a normally distributed population when the sample size is small. It is the basis of the popular Student's t-tests for the statistical significance of the difference between two sample means, and for confidence intervals for the difference between two population means.
Standard normal distribution	The standard normal distribution is the normal distribution with a mean of zero and a standard deviation of one. It is often called the bell curve because the graph of its probability density resembles a bell.
Normal distribution	In probability theory, the normal (or Gaussian) distribution is a continuous probability distribution that has a bell-shaped probability density function, known as the Gaussian function or informally the bell curve: $$f(x; \mu, \sigma^2) = \frac{1}{\sigma\sqrt{2\pi}} e^{-\frac{1}{2}\left(\frac{x-\mu}{\sigma}\right)^2}$$ The parameter µ is the mean or expectation (location of the peak) and σ^2 is the variance. σ is known as the standard deviation. The distribution with µ = 0 and

Chapter 8. Statistical Inference: Confidence Intervals

Financial statement	A Financial statement is a formal record of the financial activities of a business, person, or other entity. In British English--including United Kingdom company law--a Financial statement is often referred to as an account, although the term Financial statement is also used, particularly by accountants.
	For a business enterprise, all the relevant financial information, presented in a structured manner and in a form easy to understand, are called the Financial statements. They typically include four basic Financial statements:
	· Balance sheet: also referred to as statement of financial position or condition, reports on a company's assets, liabilities, and Ownership equity at a given point in time. · Income statement: also referred to as Profit and Loss statement , reports on a company's income, expenses, and profits over a period of time.
Outlier	In statistics, an outlier is an observation that is numerically distant from the rest of the data. Grubbs defined an outlier as:'
	An outlying observation, or outlier, is one that appears to deviate markedly from other members of the sample in which it occurs. '
	Outliers can occur by chance in any distribution, but they are often indicative either of measurement error or that the population has a heavy-tailed distribution.
Relative standing	Relative standing is a measurement of numbers which indicate where a particular values lies in relation to the rest of the values in a set of data or population.
Gosset	In 1994, the house was sold to Rémy-Cointreau. Under the management of Beatrice Cointreau, (also the head of Cognac Frapin), Gosset succeeded in increasing its production to one million bottles in 2005.
	Gosset wines are made from a blend of Pinot Noir, Chardonnay, and Pinot Meunier.
P-value	In statistical significance testing, the p-value is the probability of obtaining a test statistic at least as extreme as the one that was actually observed, assuming that the null hypothesis is true. In this context, value a is considered more 'extreme' than b if a is less likely to occur under the null. One often 'rejects the null hypothesis' when the p-value is less than the significance level α, which is often 0.05 or 0.01. When the null hypothesis is rejected, the result is said to be statistically significant.
Estimation	In project management (i.e., for engineering), accurate estimates are the basis of sound project planning.

Many processes have been developed to aid engineers in making accurate estimates, such as•Analogy based estimation•Compartmentalization (i.e., breakdown of tasks)•Delphi method•Documenting estimation results•Educated assumptions•Estimating each task•Examining historical data•Identifying dependencies•Parametric estimating•Risk assessment•Structured planning

Popular estimation processes for software projects include:•Cocomo•Cosysmo•Event chain methodology•Function points•Program Evaluation and Review Technique (PERT)•Proxy Based Estimation (PROBE) (from the Personal Software Process)•The Planning Game (from Extreme Programming)•Weighted Micro Function Points (WMFP)•Wideband Delphi.

Exact test	In statistics, an exact (significance) test is a test where all assumptions upon which the derivation of the distribution of the test statistic is based are met, as opposed to an approximate test, in which the approximation may be made as close as desired by making the sample size big enough. This will result in a significance test that will have a false rejection rate always equal to the significance level of the test. For example an exact test at significance level 5% will in the long run reject true null hypothesis exactly 5% of the time.
ANOVA	In statistics, ANOVA is a collection of statistical models, and their associated procedures, in which the observed variance is partitioned into components due to different sources of variation. In its simplest form ANOVA provides a statistical test of whether or not the means of several groups are all equal, and therefore generalizes Student's two-sample t-test to more than two groups. ANOVAs are helpful because they possess a certain advantage over a two-sample t-test. Doing multiple two-sample t-tests would result in a largely increased chance of committing a type I error. For this reason, ANOVAs are useful in comparing three or more means. There are three conceptual classes of such models: · Fixed-effects models assume that the data came from normal populations which may differ only in their means. (Model 1) · Random effects models assume that the data describe a hierarchy of different populations whose differences are constrained by the hierarchy. (Model 2) · Mixed-effect models describe the situations where both fixed and random effects are present. (Model 3)
Factorization	In mathematics, Factorization or factoring is the decomposition of an object into a product of other objects, which when multiplied together give the original. For example, the number 15 factors into primes as 3×5, and the polynomial $x^2 - 4$ factors as $(x - 2)(x + 2)$. In all cases, a product of simpler objects is obtained.
Ronald Aylmer Fisher	Sir Ronald Aylmer Fisher FRS (17 February 1890 - 29 July 1962) was an English statistician, evolutionary biologist, eugenicist and geneticist.

Chapter 8. Statistical Inference: Confidence Intervals

1. A _____ is a formal record of the financial activities of a business, person, or other entity. In British English--including United Kingdom company law--a _____ is often referred to as an account, although the term _____ is also used, particularly by accountants.

   For a business enterprise, all the relevant financial information, presented in a structured manner and in a form easy to understand, are called the _____s. They typically include four basic _____s:

   · Balance sheet: also referred to as statement of financial position or condition, reports on a company's assets, liabilities, and Ownership equity at a given point in time. · Income statement: also referred to as Profit and Loss statement , reports on a company's income, expenses, and profits over a period of time.

   a. Cherenkov radiation
   b. Cyclotron radiation
   c. Synchrotron radiation
   d. Financial statement

2. A _____, in performing arts, generally comprises an event in which one group of people (the performer or performers) behave in a particular way for another group of people (the audience). Sometimes the dividing line between performer and the audience may become blurred, as in the example of 'participatory theatre' where audience members might get involved in the production. Singing choral music, and performing in a ballet are examples.

   a. 1-factor
   b. Wastebasket taxon
   c. Web-based taxonomy
   d. Performance

3. _____ is an estimator that has its expected value the parametric value. An estimator or decision rule having nonzero bias is said to be biased.

   a. A priori probability
   b. Extinction probability
   c. Automated reasoning
   d. Unbiased estimator

4. In 1994, the house was sold to Rémy-Cointreau. Under the management of Beatrice Cointreau, (also the head of Cognac Frapin), _____ succeeded in increasing its production to one million bottles in 2005.

   _____ wines are made from a blend of Pinot Noir, Chardonnay, and Pinot Meunier.

   a. 1-factor
   b. Random error
   c. Morbidity rate
   d. Gosset

5. In statistics, an _____ is a rule for calculating an estimate of a given quantity based on observed data: thus the rule and its result (the estimate) are distinguished.

   There are point and interval _____s. The point _____s yield single-valued results, although this includes the possibility of single vector-valued results and results that can be expressed as a single function.

   a. Estimator
   b. Inductive reasoning
   c. Interval estimation
   d. Inverse probability

1. d
2. d
3. d
4. d
5. a

You can take the complete Chapter Practice Test

for Chapter 8. Statistical Inference: Confidence Intervals
on all key terms, persons, places, and concepts.

Online 99 Cents

http://www.epub27.14.20551.8.cram101.com/

Use www.Cram101.com for all your study needs

including Cram101's online interactive problem solving labs in

chemistry, statistics, mathematics, and more.

CHAPTER OUTLINE: KEY TERMS, PEOPLE, PLACES, CONCEPTS

	Alternative hypothesis
	Null hypothesis
	P-value
	Standard error
	Test statistic
	Statistic
	Parameter
	Standard normal distribution
	Normal distribution
	Statistically significant
	Statistical significance
	Confidence interval
	Population mean
	Binomial
	Binomial test
	ANOVA
	Chi-squared test
	Relative standing
	Type I error

	Type II error
	Practical significance
	Data analysis
	Jerzy Neyman
	Changing-criterion research design
	Likelihood
	Single event upset
	Performance

| Alternative hypothesis | In statistical hypothesis testing, the alternative hypothesis and the null hypothesis are the two rival hypotheses which are compared by a statistical hypothesis test. An example might be where water quality in a stream has been observed over many years and a test is made of the null hypothesis that there is no change in quality between the first and second halves of the data against the alternative hypothesis that the quality is poorer in the second half of the record.

In the case of a scalar parameter, there are four principal types of alternative hypothesis: a point alternative hypothesis, a one-tailed directional alternative hypothesis, a two-tailed directional alternative hypothesis, and an non-directional alternative hypothesis. |
| --- | --- |
| Null hypothesis | The practice of science involves formulating and testing hypotheses, assertions that are falsifiable using a test of observed data. The null hypothesis typically corresponds to a general or default position. For example, there is no relationship between two measured phenomena, or a potential treatment has no effect. |

P-value	In statistical significance testing, the p-value is the probability of obtaining a test statistic at least as extreme as the one that was actually observed, assuming that the null hypothesis is true. In this context, value a is considered more 'extreme' than b if a is less likely to occur under the null. One often 'rejects the null hypothesis' when the p-value is less than the significance level α, which is often 0.05 or 0.01. When the null hypothesis is rejected, the result is said to be statistically significant.
Standard error	The standard error is the standard deviation of the sampling distribution of a statistic. The term may also be used to refer to an estimate of that standard deviation, derived from a particular sample used to compute the estimate.

For example, the sample mean is the usual estimator of a population mean. |
| Test statistic | In statistical hypothesis testing, a hypothesis test is typically specified in terms of a test statistic, which is a function of the sample; it is considered as a numerical summary of a set of data that reduces the data to one or a small number of values that can be used to perform a hypothesis test. Given a null hypothesis and a test statistic T, we can specify a 'null value' T_0 such that values of T close to T_0 present the strongest evidence in favor of the null hypothesis, whereas values of T far from T_0 present the strongest evidence against the null hypothesis. An important property of a test statistic is that we must be able to determine its sampling distribution under the null hypothesis, which allows us to calculate p-values. |
| Statistic | A statistic is a single measure of some attribute of a sample (e.g. its arithmetic mean value). It is calculated by applying a function (statistical algorithm) to the values of the items comprising the sample which are known together as a set of data.

More formally, statistical theory defines a statistic as a function of a sample where the function itself is independent of the sample's distribution; that is, the function can be stated before realisation of the data. |
| Parameter | Parameter can be interpreted in mathematics, logic, linguistics, environmental science and other disciplines.

In its common meaning, the term is used to identify a characteristic, a feature, a measurable factor that can help in defining a particular system. It is an important element to take into consideration for the evaluation or for the comprehension of an event, a project or any situation. |
| Standard normal distribution | The standard normal distribution is the normal distribution with a mean of zero and a standard deviation of one. It is often called the bell curve because the graph of its probability density resembles a bell. |

Chapter 9. Statistical Inference: Significance Tests About Hypotheses

Normal distribution	In probability theory, the normal (or Gaussian) distribution is a continuous probability distribution that has a bell-shaped probability density function, known as the Gaussian function or informally the bell curve: $$f(x; \mu, \sigma^2) = \frac{1}{\sigma\sqrt{2\pi}} e^{-\frac{1}{2}\left(\frac{x-\mu}{\sigma}\right)^2}$$ The parameter μ is the mean or expectation (location of the peak) and σ^2 is the variance. σ is known as the standard deviation. The distribution with μ = 0 and σ^2 = 1 is called the standard normal distribution or the unit normal distribution.
Statistically significant	In statistics, a result is called statistically significant if it is unlikely to have occurred by chance. The phrase test of significance was coined by Ronald Fisher. The use of the word significance in statistics is different from the standard one, which suggests that something is important or meaningful.
Statistical significance	In statistics, a result is called 'statistically significant' if it is unlikely to have occurred by chance. The phrase test of significance was coined by Ronald Fisher. As used in statistics, significant does not mean important or meaningful, as it does in everyday speech. Research analysts who focus solely on significant results may miss important response patterns which individually may fall under the threshold set for tests of significance. Many researchers urge that tests of significance should always be accompanied by effect-size statistics, which approximate the size and thus the practical importance of the difference. The amount of evidence required to accept that an event is unlikely to have arisen by chance is known as the significance level or critical p-value: in traditional Fisherian statistical hypothesis testing, the p-value is the probability of observing data at least as extreme as that observed, given that the null hypothesis is true. If the obtained p-value is small then it can be said either the null hypothesis is false or an unusual event has occurred. P-values do not have any repeat sampling interpretation. An alternative (but nevertheless related) statistical hypothesis testing framework is the Neyman-Pearson frequentist school which requires both a null and an alternative hypothesis to be defined and investigates the repeat sampling properties of the procedure, i.e. the probability that a decision to reject the null hypothesis will be made when it is in fact true and should not have been rejected (this is called a 'false positive' or Type I error) and the probability that a decision will be made to accept the null hypothesis when it is in fact false (Type II error). Fisherian p-values are philosophically different from Neyman-Pearson Type I errors. This confusion is unfortunately propagated by many statistics textbooks. Use in practice The significance level is usually denoted by the Greek symbol α (lowercase alpha).

Popular levels of significance are 10% (0.1), 5% (0.05), 1% (0.01), 0.5% (0.005), and 0.1% (0.001). If a test of significance gives a p-value lower than the significance level α, the null hypothesis is rejected. Such results are informally referred to as 'statistically significant'. For example, if someone argues that 'there's only one chance in a thousand this could have happened by coincidence,' a 0.001 level of statistical significance is being implied.

Confidence interval	In statistics, a confidence interval is a kind of interval estimate of a population parameter and is used to indicate the reliability of an estimate. It is an observed interval (i.e. it is calculated from the observations), in principle different from sample to sample, that frequently includes the parameter of interest, if the experiment is repeated. How frequently the observed interval contains the parameter is determined by the confidence level or confidence coefficient.
Population mean	The mean of a population has an expected value of μ, known as the population mean. The sample mean makes a good estimator of the population mean, as its expected value is the same as the population mean. The sample mean of a population is a random variable, not a constant, and consequently it will have its own distribution. For a random sample of n observations from a normally distributed population, the sample mean distribution is $$\bar{x} \sim N \left\{ \mu, \frac{\sigma^2}{n} \right\}.$$ Often, since the population variance is an unknown parameter, it is estimated by the mean sum of squares, which changes the distribution of the sample mean from a normal distribution to a Student's t distribution with n − 1 degrees of freedom.
Binomial	In elementary algebra, a Binomial is a polynomial with two terms--the sum of two monomials--often bound by parenthesis or brackets when operated upon. It is the simplest kind of polynomial other than monomials. · The Binomial $a^2 - b^2$ can be factored as the product of two other Binomials: $a^2 - b^2 = (a + b)(a - b)$. This is a special case of the more general formula: $$a^{n+1} - b^{n+1} = (a - b) \sum_{k=0}^{n} a^k b^{n-k}.$$ · The product of a pair of linear Binomials (ax + b) and (cx + d) is:

$(ax + b)(cx + d) = acx^2 + axd + bcx + bd$.

· A Binomial raised to the nth power, represented as

$(a + b)^n$

can be expanded by means of the Binomial theorem or, equivalently, using Pascal's triangle. Taking a simple example, the perfect square Binomial $(p + q)^2$ can be found by squaring the first term, adding twice the product of the first and second terms and finally adding the square of the second term, to give $p^2 + 2pq + q^2$.

· A simple but interesting application of the cited Binomial formula is the '(m,n)-formula' for generating Pythagorean triples: for $m < n$, let $a = n^2 - m^2$, $b = 2mn$, $c = n^2 + m^2$, then $a^2 + b^2 = c^2$.

| Binomial test | In statistics, the binomial test is an exact test of the statistical significance of deviations from a theoretically expected distribution of observations into two categories.

The most common use of the binomial test is in the case where the null hypothesis is that two categories are equally likely to occur (such as a coin toss). Tables are widely available to give the significance observed numbers of observations in the categories for this case. |

| ANOVA | In statistics, ANOVA is a collection of statistical models, and their associated procedures, in which the observed variance is partitioned into components due to different sources of variation. In its simplest form ANOVA provides a statistical test of whether or not the means of several groups are all equal, and therefore generalizes Student's two-sample t-test to more than two groups. ANOVAs are helpful because they possess a certain advantage over a two-sample t-test. Doing multiple two-sample t-tests would result in a largely increased chance of committing a type I error. For this reason, ANOVAs are useful in comparing three or more means.

There are three conceptual classes of such models:

· Fixed-effects models assume that the data came from normal populations which may differ only in their means. (Model 1) · Random effects models assume that the data describe a hierarchy of different populations whose differences are constrained by the hierarchy. (Model 2) · Mixed-effect models describe the situations where both fixed and random effects are present. |

Chi-squared test	A chi-squared test, also referred to as chi-square test or χ^2 test, is any statistical hypothesis test in which the sampling distribution of the test statistic is a chi-squared distribution when the null hypothesis is true, or any in which this is asymptotically true, meaning that the sampling distribution (if the null hypothesis is true) can be made to approximate a chi-squared distribution as closely as desired by making the sample size large enough.
	Some examples of chi-squared tests where the chi-squared distribution is only approximately valid:•Pearson's chi-squared test, also known as the chi-squared goodness-of-fit test or chi-squared test for independence. When mentioned without any modifiers or without other precluding context, this test is usually understood .•Yates's correction for continuity, also known as Yates' chi-squared test.•Cochran-Mantel-Haenszel chi-squared test.•McNemar's test, used in certain 2 × 2 tables with pairing•Linear-by-linear association chi-squared test•The portmanteau test in time-series analysis, testing for the presence of autocorrelation•Likelihood-ratio tests in general statistical modelling, for testing whether there is evidence of the need to move from a simple model to a more complicated one (where the simple model is nested within the complicated one).
	One case where the distribution of the test statistic is an exact chi-squared distribution is the test that the variance of a normally distributed population has a given value based on a sample variance.
Relative standing	Relative standing is a measurement of numbers which indicate where a particular values lies in relation to the rest of the values in a set of data or population.
Type I error	In statistics, the terms Type I error (also, α error, false alarm rate (FAR) or false positive) and type II error (β error) are used to describe possible errors made in a statistical decision process. In 1928, Jerzy Neyman (1894-1981) and Egon Pearson (1895-1980), both eminent statisticians, discussed the problems associated with 'deciding whether or not a particular sample may be judged as likely to have been randomly drawn from a certain population' (1928/1967, p.1): and identified 'two sources of error', namely:
	Type I (α): reject the null hypothesis when the null hypothesis is true, and
	Type II (β): fail to reject the null hypothesis when the null hypothesis is false
	In 1930, they elaborated on these two sources of error, remarking that 'in testing hypotheses two considerations must be kept in view, (1) we must be able to reduce the chance of rejecting a true hypothesis to as low a value as desired; (2) the test must be so devised that it will reject the hypothesis tested when it is likely to be false.'
	Scientists recognize two different sorts of error:

· Statistical error: the difference between a computed, estimated, or measured value and the true, specified, or theoretically correct value that is caused by random, and inherently unpredictable fluctuations in the measurement apparatus or the system being studied. · Systematic error: the difference between a computed, estimated, or measured value and the true, specified, or theoretically correct value that is caused by non-random fluctuations from an unknown source , and which, once identified, can usually be eliminated.

Statisticians speak of two significant sorts of statistical error. The context is that there is a 'null hypothesis' which corresponds to a presumed default 'state of nature', e.g., that an individual is free of disease, that an accused is innocent, or that a potential login candidate is not authorized.

Type II error	In statistics, the terms type I error (also, α error, false alarm rate (FAR) or false positive) and Type II error (β error) are used to describe possible errors made in a statistical decision process. In 1928, Jerzy Neyman (1894-1981) and Egon Pearson (1895-1980), both eminent statisticians, discussed the problems associated with 'deciding whether or not a particular sample may be judged as likely to have been randomly drawn from a certain population' (1928/1967, p.1): and identified 'two sources of error', namely:

Type I (α): reject the null hypothesis when the null hypothesis is true, and

Type II (β): fail to reject the null hypothesis when the null hypothesis is false

In 1930, they elaborated on these two sources of error, remarking that 'in testing hypotheses two considerations must be kept in view, (1) we must be able to reduce the chance of rejecting a true hypothesis to as low a value as desired; (2) the test must be so devised that it will reject the hypothesis tested when it is likely to be false.'

Scientists recognize two different sorts of error:

· Statistical error: the difference between a computed, estimated, or measured value and the true, specified, or theoretically correct value that is caused by random, and inherently unpredictable fluctuations in the measurement apparatus or the system being studied. · Systematic error: the difference between a computed, estimated, or measured value and the true, specified, or theoretically correct value that is caused by non-random fluctuations from an unknown source , and which, once identified, can usually be eliminated.

Statisticians speak of two significant sorts of statistical error. The context is that there is a 'null hypothesis' which corresponds to a presumed default 'state of nature', e.g., that an individual is free of disease, that an accused is innocent, or that a potential login candidate is not authorized.

Practical significance	A common misconception is that a statistically significant result is always of practical significance, or demonstrates a large effect in the population. Given a sufficiently large sample, extremely small and non-notable differences can be found to be statistically significant, and statistical significance says nothing about the practical significance of a difference.
Data analysis	Analysis of data is a process of inspecting, cleaning, transforming, and modeling data with the goal of highlighting useful information, suggesting conclusions, and supporting decision making. Data analysis has multiple facets and approaches, encompassing diverse techniques under a variety of names, in different business, science, and social science domains.
	Data mining is a particular data analysis technique that focuses on modeling and knowledge discovery for predictive rather than purely descriptive purposes.
Jerzy Neyman	Jerzy Neyman born Jerzy Splawa-Neyman, was a Polish American mathematician and statistician who spent most of his professional career at the University of California, Berkeley.
	He was born into a Polish family in Bendery, Bessarabia in Imperial Russia, the fourth of four children of Czeslaw Splawa-Neyman and Kazimiera Lutoslawska. His family was Roman Catholic and Neyman served as an altar boy during his early childhood.
Changing-criterion research design	In a changing-criterion research design a criterion for reinforcement is changed across the experiment to demonstrate the functional relationship between the reinforcement and the behavior. See Mark Dixon's work with a participant using a short video clip to generate a preference for a progressively delayed variable reinforcement over a fixed shorter delay reinforcement in physical therapy.
Likelihood	In statistics, the Likelihood function (often simply the Likelihood) is a function of the parameters of a statistical model that plays a key role in statistical inference. In non-technical parlance, 'Likelihood' is usually a synonym for 'probability', but in statistical usage there is a clear distinction: whereas 'probability' allows us to predict unknown outcomes based on known parameters, 'Likelihood' allows us to estimate unknown parameters based on known outcomes.
	In a sense, Likelihood can be thought a reversed version of conditional probability.
Single event upset	A single event upset is a change of state caused by ions or electro-magnetic radiation striking a sensitive node in a micro-electronic device, such as in a microprocessor, semiconductor memory). The error in device output or operation caused as a result of the strike is called an single event upset or a soft error.

Chapter 9. Statistical Inference: Significance Tests About Hypotheses

Performance	A performance, in performing arts, generally comprises an event in which one group of people (the performer or performers) behave in a particular way for another group of people (the audience). Sometimes the dividing line between performer and the audience may become blurred, as in the example of 'participatory theatre' where audience members might get involved in the production. Singing choral music, and performing in a ballet are examples.

1. In statistics, the _____ is an exact test of the statistical significance of deviations from a theoretically expected distribution of observations into two categories.

   The most common use of the _____ is in the case where the null hypothesis is that two categories are equally likely to occur (such as a coin toss). Tables are widely available to give the significance observed numbers of observations in the categories for this case.

   a. Checking whether a coin is fair
   b. Binomial test
   c. Chow test
   d. Closed testing procedure

2. _____ can be interpreted in mathematics, logic, linguistics, environmental science and other disciplines.

   In its common meaning, the term is used to identify a characteristic, a feature, a measurable factor that can help in defining a particular system. It is an important element to take into consideration for the evaluation or for the comprehension of an event, a project or any situation.

   a. Parts-per notation
   b. Pathological
   c. Parameter
   d. Plugging in

3. . In statistical hypothesis testing, the _____ and the null hypothesis are the two rival hypotheses which are compared by a statistical hypothesis test. An example might be where water quality in a stream has been observed over many years and a test is made of the null hypothesis that there is no change in quality between the first and second halves of the data against the _____ that the quality is poorer in the second half of the record.

   In the case of a scalar parameter, there are four principal types of _____: a point _____, a one-tailed directional _____, a two-tailed directional _____, and an non-directional _____.

   a. Exact statistics

b. bivariate hyperbolic operator of the second order

c. Blaine Lawson

d. Alternative hypothesis

4. In a _____ a criterion for reinforcement is changed across the experiment to demonstrate the functional relationship between the reinforcement and the behavior. See Mark Dixon's work with a participant using a short video clip to generate a preference for a progressively delayed variable reinforcement over a fixed shorter delay reinforcement in physical therapy.

a. Habitual offender

b. Changing-criterion research design

c. Test panel

d. Package testing

5. The practice of science involves formulating and testing hypotheses, assertions that are falsifiable using a test of observed data. The _____ typically corresponds to a general or default position. For example, there is no relationship between two measured phenomena, or a potential treatment has no effect.

a. Parametric statistics

b. Foundations of statistics

c. Null hypothesis

d. Philosophy of statistics

1. b
2. c
3. d
4. b
5. c

You can take the complete Chapter Practice Test

for Chapter 9. Statistical Inference: Significance Tests About Hypotheses
on all key terms, persons, places, and concepts.

Online 99 Cents

http://www.epub27.14.20551.9.cram101.com/

Use www.Cram101.com for all your study needs

including Cram101's online interactive problem solving labs in

chemistry, statistics, mathematics, and more.

Chapter 10. Comparing Two Groups

CHAPTER OUTLINE: KEY TERMS, PEOPLE, PLACES, CONCEPTS

Financial statement

Binary variable

Response variable

Parameter

Standard error

Confidence interval

Small-sample

Inference

Population mean

Cell

Pooled standard deviation

Standard deviation

Deviation

Ratio

Relative risk

ANOVA

T-test

Reaction time

Belief

CHAPTER OUTLINE: KEY TERMS, PEOPLE, PLACES, CONCEPTS

	Speech recognition
	Control variable
	Paradox
	Null hypothesis
	P-value
	Statistically significant
	Test statistic
	Type I error
	Type II error
	Statistic
	Student t distribution
	Margin of error
	Randomization
	Sampling distribution

Chapter 10. Comparing Two Groups

Financial statement	A Financial statement is a formal record of the financial activities of a business, person, or other entity. In British English--including United Kingdom company law--a Financial statement is often referred to as an account, although the term Financial statement is also used, particularly by accountants.
	For a business enterprise, all the relevant financial information, presented in a structured manner and in a form easy to understand, are called the Financial statements. They typically include four basic Financial statements:
	· Balance sheet: also referred to as statement of financial position or condition, reports on a company's assets, liabilities, and Ownership equity at a given point in time. · Income statement: also referred to as Profit and Loss statement , reports on a company's income, expenses, and profits over a period of time.
Binary variable	The term binary data has various meanings in different technical fields. In general, it refers to a unit of data which can take on only two possible values, traditionally termed 0 and 1 in accordance with the binary numeral system. Related concepts in various fields are *logical value in logic, which represents the truth or falsehood of a logical proposition•Boolean value, a representation of the concepts 'true' or 'false' used to do Boolean arithmetic in logic and computer science•binary digit, a single 0 or 1 in a binary number, used to represent numbers in base 2 (the binary numeral system) In statistics
	In statistics, binary data is a statistical data type described by binary variables, which can take only two possible values.
Response variable	The terms 'dependent variable' and 'independent variable' are used in similar but subtly different ways in mathematics and statistics as part of the standard terminology in those subjects. They are used to distinguish between two types of quantities being considered, separating them into those available at the start of a process and those being created by it, where the latter (dependent variables) are dependent on the former (independent variables).
	The independent variable is typically the variable representing the value being manipulated or changed and the dependent variable is the observed result of the independent variable being manipulated. For example concerning nutrition, the independent variable of daily vitamin C intake (how much vitamin C one consumes) can influence the dependent variable of life expectancy (the average age one attains). Over some period of time, scientists will control the vitamin C intake in a substantial group of people. One part of the group will be given a daily high dose of vitamin C, and the remainder will be given a placebo pill (so that they are unaware of not belonging to the first group) without vitamin C. The scientists will investigate if there is any statistically significant difference in the life span of the people who took the high dose and those who took the placebo (no dose).

The goal is to see if the independent variable of high vitamin C dosage has a correlation with the dependent variable of people's life span. The designation independent/dependent is clear in this case, because if a correlation is found, it cannot be that life span has influenced vitamin C intake, but an influence in the other direction is possible. Use in mathematics

In calculus, a function is a map whose action is specified on variables. Take x and y to be two variables. A function f may map x to some expression in x. Assigning $y = f(x)$ gives a relation between x and y. If there is some relation specifying y in terms of x, then y is known as a 'dependent variable' (and x is an 'independent variable'). Use in statistics Controlled experiments

In a statistics experiment, the dependent variable is the event studied and expected to change whenever the independent variable is altered.

In the design of experiments, an independent variable's values are controlled or selected by the experimenter to determine its relationship to an observed phenomenon (i.e., the dependent variable). In such an experiment, an attempt is made to find evidence that the values of the independent variable determine the values of the dependent variable. The independent variable can be changed as required, and its values do not represent a problem requiring explanation in an analysis, but are taken simply as given. The dependent variable, on the other hand, usually cannot be directly controlled.

Controlled variables are also important to identify in experiments. They are the variables that are kept constant to prevent their influence on the effect of the independent variable on the dependent. Every experiment has a controlling variable, and it is necessary to not change it, or the results of the experiment won't be valid.

'Extraneous variables' are those that might affect the relationship between the independent and dependent variables. Extraneous variables are usually not theoretically interesting. They are measured in order for the experimenter to compensate for them. For example, an experimenter who wishes to measure the degree to which caffeine intake (the independent variable) influences explicit recall for a word list (the dependent variable) might also measure the participant's age (extraneous variable). She can then use these age data to control for the uninteresting effect of age, clarifying the relationship between caffeine and memory.

In summary:•Independent variables answer the question 'What do I change?'•Dependent variables answer the question 'What do I observe?'•Controlled variables answer the question 'What do I keep the same?'•Extraneous variables answer the question 'What uninteresting variables might mediate the effect of the IV on the DV?'Alternative terminology in statistics

Chapter 10. Comparing Two Groups

	In statistics, the dependent/independent variable terminology is used more widely than just in relation to controlled experiments. For example the data analysis of two jointly varying quantities may involve treating each in turn as the dependent variable and the other as the independent variable. However, for general usage, the pair response variable and explanatory variable is preferable as quantities treated as 'independent variables' are rarely statistically independent.
Parameter	Parameter can be interpreted in mathematics, logic, linguistics, environmental science and other disciplines. In its common meaning, the term is used to identify a characteristic, a feature, a measurable factor that can help in defining a particular system. It is an important element to take into consideration for the evaluation or for the comprehension of an event, a project or any situation.
Standard error	The standard error is the standard deviation of the sampling distribution of a statistic. The term may also be used to refer to an estimate of that standard deviation, derived from a particular sample used to compute the estimate. For example, the sample mean is the usual estimator of a population mean.
Confidence interval	In statistics, a confidence interval is a kind of interval estimate of a population parameter and is used to indicate the reliability of an estimate. It is an observed interval (i.e. it is calculated from the observations), in principle different from sample to sample, that frequently includes the parameter of interest, if the experiment is repeated. How frequently the observed interval contains the parameter is determined by the confidence level or confidence coefficient.
Small-sample	Small-sample correction is a correction to the information or the uncertainty measure to account for this effect. In terms of statistics, the uncertainty measure is biased when there are small number of samples.
Inference	Inference is the act of drawing a conclusion by deductive reasoning from given facts. The conclusion drawn is also called an inference. The laws of valid inference are studied in the field of logic.
Population mean	The mean of a population has an expected value of μ, known as the population mean. The sample mean makes a good estimator of the population mean, as its expected value is the same as the population mean. The sample mean of a population is a random variable, not a constant, and consequently it will have its own distribution. For a random sample of n observations from a normally distributed population, the sample mean distribution is

$$\bar{x} \sim N \left\{ \mu, \frac{\sigma^2}{n} \right\}.$$

Often, since the population variance is an unknown parameter, it is estimated by the mean sum of squares, which changes the distribution of the sample mean from a normal distribution to a Student's t distribution with n − 1 degrees of freedom.

Cell	In geometry, a cell is a three-dimensional element that is part of a higher-dimensional object.
	In polytopes
	A cell is a three-dimensional polyhedron element that is part of the boundary of a higher-dimensional polytope, such as a polychoron (4-polytope) or honeycomb (3-space tessellation).
	For example, a cubic honeycomb is made of cubic cells, with 4 cubes on each edge.
Pooled standard deviation	In statistics, Pooled standard deviation is a way to find an estimate of the population standard deviation given several different samples taken in different circumstances where the mean may vary between samples but the true standard deviation (precision) is assumed to remain the same.

It is calculated by

$$s_p = \sqrt{\frac{\sum_{i=1}^{k}((n_i - 1)s_i^2)}{\sum_{i=1}^{k}(n_i - 1)}}$$

or with simpler notation,

$$s_p = \sqrt{\frac{(n_1 - 1)s_1^2 + (n_2 - 1)s_2^2 + \cdots + (n_k - 1)s_k^2}{n_1 + n_2 + \cdots + n_k - k}}$$

where s_p is the Pooled standard deviation, n_i is the sample size of the i'th sample, s_i is the standard deviation of the ith sample, and k is the number of samples being combined. n − 1 is used instead of n for the same reason it may be used in calculating standard deviations from samples.

Standard deviation	Standard deviation is a widely used measurement of variability or diversity used in statistics and probability theory. It shows how much variation or 'dispersion' there is from the 'average' (mean, or expected/budgeted value). A low standard deviation indicates that the data points tend to be very close to the mean, whereas high standard deviation indicates that the data are spread out over a large range of values.
Deviation	In mathematics and statistics, deviation is a measure of difference between the observed value and the mean. The sign of deviation (positive or negative), reports the direction of that difference (it is larger when the sign is positive, and smaller if it is negative).

Chapter 10. Comparing Two Groups

Ratio	A ratio is an expression that compares quantities relative to each other. The most common examples involve two quantities, but any number of quantities can be compared. ratios are represented mathematically by separating each quantity with a colon - for example, the ratio 2:3, which is read as the ratio 'two to three'.
Relative risk	In statistics and mathematical epidemiology, relative risk is the risk of an event relative to exposure. Relative risk is a ratio of the probability of the event occurring in the exposed group versus a non-exposed group. $$RR = \frac{p_{\text{exposed}}}{p_{\text{non-exposed}}}$$ Consider an example where the probability of developing lung cancer among smokers was 20% and among non-smokers 1%.
ANOVA	In statistics, ANOVA is a collection of statistical models, and their associated procedures, in which the observed variance is partitioned into components due to different sources of variation. In its simplest form ANOVA provides a statistical test of whether or not the means of several groups are all equal, and therefore generalizes Student's two-sample t-test to more than two groups. ANOVAs are helpful because they possess a certain advantage over a two-sample t-test. Doing multiple two-sample t-tests would result in a largely increased chance of committing a type I error. For this reason, ANOVAs are useful in comparing three or more means.

There are three conceptual classes of such models:

· Fixed-effects models assume that the data came from normal populations which may differ only in their means. (Model 1) · Random effects models assume that the data describe a hierarchy of different populations whose differences are constrained by the hierarchy. (Model 2) · Mixed-effect models describe the situations where both fixed and random effects are present. (Model 3) |
| T-test | A t-test is any statistical hypothesis test in which the test statistic follows a Student's t distribution if the null hypothesis is true. It is most commonly applied when the test statistic would follow a normal distribution if the value of a scaling term in the test statistic were known. When the scaling term is unknown and is replaced by an estimate based on the data, the test statistic (under certain conditions) follows a Student's t distribution. |
| Reaction time | Reaction time is the elapsed time between the presentation of a sensory stimulus and the subsequent behavioral response. It is often used in experimental psychology to measure the duration of mental operations, known as mental chronometry. |

CHAPTER HIGHLIGHTS & NOTES: KEY TERMS, PEOPLE, PLACES, CONCEPTS

Belief	Belief is the psychological state in which an individual holds a proposition or premise to be true.
	The terms Belief and knowledge are used differently in philosophy.
	Epistemology is the philosophical study of knowledge and Belief.
Speech recognition	Speech recognition converts spoken words to text. The term 'voice recognition' is sometimes used to refer to recognition systems that must be trained to a particular speaker--as is the case for most desktop recognition software. Recognizing the speaker can simplify the task of translating speech.
Control variable	The term control variable has different meanings, depending on the area/place in which it is used. The control variable is something that is constant and unchanged in an experiment.
	A control variable is any factor that remains unchanged and strongly influences values; it is held constant to test the relative impact of an independent variable.
Paradox	A Paradox is a true statement or group of statements that leads to a contradiction or a situation which defies intuition. The term is also used for an apparent contradiction that actually expresses a non-dual truth (cf. kÅan, Catuskoti).
Null hypothesis	The practice of science involves formulating and testing hypotheses, assertions that are falsifiable using a test of observed data. The null hypothesis typically corresponds to a general or default position. For example, there is no relationship between two measured phenomena, or a potential treatment has no effect.
P-value	In statistical significance testing, the p-value is the probability of obtaining a test statistic at least as extreme as the one that was actually observed, assuming that the null hypothesis is true. In this context, value a is considered more 'extreme' than b if a is less likely to occur under the null. One often 'rejects the null hypothesis' when the p-value is less than the significance level α, which is often 0.05 or 0.01. When the null hypothesis is rejected, the result is said to be statistically significant.
Statistically significant	In statistics, a result is called statistically significant if it is unlikely to have occurred by chance. The phrase test of significance was coined by Ronald Fisher.
	The use of the word significance in statistics is different from the standard one, which suggests that something is important or meaningful.

Chapter 10. Comparing Two Groups

Test statistic	In statistical hypothesis testing, a hypothesis test is typically specified in terms of a test statistic, which is a function of the sample; it is considered as a numerical summary of a set of data that reduces the data to one or a small number of values that can be used to perform a hypothesis test. Given a null hypothesis and a test statistic T, we can specify a 'null value' T_0 such that values of T close to T_0 present the strongest evidence in favor of the null hypothesis, whereas values of T far from T_0 present the strongest evidence against the null hypothesis. An important property of a test statistic is that we must be able to determine its sampling distribution under the null hypothesis, which allows us to calculate p-values.
Type I error	In statistics, the terms Type I error (also, α error, false alarm rate (FAR) or false positive) and type II error (β error) are used to describe possible errors made in a statistical decision process. In 1928, Jerzy Neyman (1894-1981) and Egon Pearson (1895-1980), both eminent statisticians, discussed the problems associated with 'deciding whether or not a particular sample may be judged as likely to have been randomly drawn from a certain population' (1928/1967, p.1): and identified 'two sources of error', namely: Type I (α): reject the null hypothesis when the null hypothesis is true, and Type II (β): fail to reject the null hypothesis when the null hypothesis is false In 1930, they elaborated on these two sources of error, remarking that 'in testing hypotheses two considerations must be kept in view, (1) we must be able to reduce the chance of rejecting a true hypothesis to as low a value as desired; (2) the test must be so devised that it will reject the hypothesis tested when it is likely to be false.' Scientists recognize two different sorts of error: · Statistical error: the difference between a computed, estimated, or measured value and the true, specified, or theoretically correct value that is caused by random, and inherently unpredictable fluctuations in the measurement apparatus or the system being studied. · Systematic error: the difference between a computed, estimated, or measured value and the true, specified, or theoretically correct value that is caused by non-random fluctuations from an unknown source , and which, once identified, can usually be eliminated. Statisticians speak of two significant sorts of statistical error. The context is that there is a 'null hypothesis' which corresponds to a presumed default 'state of nature', e.g., that an individual is free of disease, that an accused is innocent, or that a potential login candidate is not authorized.
Type II error	In statistics, the terms type I error (also, α error, false alarm rate (FAR) or false positive) and Type II error (β error) are used to describe possible errors made in a statistical decision process.

In 1928, Jerzy Neyman (1894-1981) and Egon Pearson (1895-1980), both eminent statisticians, discussed the problems associated with 'deciding whether or not a particular sample may be judged as likely to have been randomly drawn from a certain population' (1928/1967, p.1): and identified 'two sources of error', namely:

Type I (α): reject the null hypothesis when the null hypothesis is true, and

Type II (β): fail to reject the null hypothesis when the null hypothesis is false

In 1930, they elaborated on these two sources of error, remarking that 'in testing hypotheses two considerations must be kept in view, (1) we must be able to reduce the chance of rejecting a true hypothesis to as low a value as desired; (2) the test must be so devised that it will reject the hypothesis tested when it is likely to be false.'

Scientists recognize two different sorts of error:

· Statistical error: the difference between a computed, estimated, or measured value and the true, specified, or theoretically correct value that is caused by random, and inherently unpredictable fluctuations in the measurement apparatus or the system being studied. · Systematic error: the difference between a computed, estimated, or measured value and the true, specified, or theoretically correct value that is caused by non-random fluctuations from an unknown source , and which, once identified, can usually be eliminated.

Statisticians speak of two significant sorts of statistical error. The context is that there is a 'null hypothesis' which corresponds to a presumed default 'state of nature', e.g., that an individual is free of disease, that an accused is innocent, or that a potential login candidate is not authorized.

Statistic	A statistic is a single measure of some attribute of a sample (e.g. its arithmetic mean value). It is calculated by applying a function (statistical algorithm) to the values of the items comprising the sample which are known together as a set of data. More formally, statistical theory defines a statistic as a function of a sample where the function itself is independent of the sample's distribution; that is, the function can be stated before realisation of the data.
Student t distribution	The Student t distribution is a probability distribution that arises in the problem of estimating the mean of a normally distributed population when the sample size is small. It is the basis of the popular Student's t-tests for the statistical significance of the difference between two sample means, and for confidence intervals for the difference between two population means.

Chapter 10. Comparing Two Groups

Margin of error	The margin of error is a statistic expressing the amount of random sampling error in a survey's results. The larger the margin of error, the less faith one should have that the poll's reported results are close to the 'true' figures; that is, the figures for the whole population. Margin of error occurs whenever a population is incompletely sampled.
Randomization	Randomization is the process of making something random; this means:•Generating a random permutation of a sequence (such as when shuffling cards).•Selecting a random sample of a population (important in statistical sampling).•Generating random numbers.•Transforming a data stream (such as when using a scrambler in telecommunications). Applications Randomization is used in statistics and in gambling. Statistics Randomization is a core principle in statistical theory, whose importance was emphasized by Charles S. Peirce in 'Illustrations of the Logic of Science' (1877-1878) and 'A Theory of Probable Inference' (1883). Randomization-based inference is especially important in experimental design and in survey sampling.
Sampling distribution	In statistics, a sampling distribution is the probability distribution of a given statistic based on a random sample. Sampling distributions are important in statistics because they provide a major simplification on the route to statistical inference. More specifically, they allow analytical considerations to be based on the sampling distribution of a statistic, rather than on the joint probability distribution of all the individual sample values.

1. The term binary data has various meanings in different technical fields. In general, it refers to a unit of data which can take on only two possible values, traditionally termed 0 and 1 in accordance with the binary numeral system. Related concepts in various fields are *logical value in logic, which represents the truth or falsehood of a logical proposition•Boolean value, a representation of the concepts 'true' or 'false' used to do Boolean arithmetic in logic and computer science•binary digit, a single 0 or 1 in a binary number, used to represent numbers in base 2 (the binary numeral system)In statistics

   In statistics, binary data is a statistical data type described by _____s, which can take only two possible values.

   a. Bivariate data
   b. Censoring
   c. Binary variable
   d. Numerical data

2. A _____ is a formal record of the financial activities of a business, person, or other entity. In British English-- including United Kingdom company law--a _____ is often referred to as an account, although the term _____ is also used, particularly by accountants.

   For a business enterprise, all the relevant financial information, presented in a structured manner and in a form easy to understand, are called the _____s. They typically include four basic _____s:

   · Balance sheet: also referred to as statement of financial position or condition, reports on a company's assets, liabilities, and Ownership equity at a given point in time. · Income statement: also referred to as Profit and Loss statement , reports on a company's income, expenses, and profits over a period of time.

   a. Cherenkov radiation
   b. Cyclotron radiation
   c. Synchrotron radiation
   d. Financial statement

3. . The terms 'dependent variable' and 'independent variable' are used in similar but subtly different ways in mathematics and statistics as part of the standard terminology in those subjects. They are used to distinguish between two types of quantities being considered, separating them into those available at the start of a process and those being created by it, where the latter (dependent variables) are dependent on the former (independent variables).

   The independent variable is typically the variable representing the value being manipulated or changed and the dependent variable is the observed result of the independent variable being manipulated. For example concerning nutrition, the independent variable of daily vitamin C intake (how much vitamin C one consumes) can influence the dependent variable of life expectancy (the average age one attains). Over some period of time, scientists will control the vitamin C intake in a substantial group of people. One part of the group will be given a daily high dose of vitamin C, and the remainder will be given a placebo pill (so that they are unaware of not belonging to the first group) without vitamin C. The scientists will investigate if there is any statistically significant difference in the life span of the people who took the high dose and those who took the placebo (no dose). The goal is to see if the independent variable of high vitamin C dosage has a correlation with the dependent variable of people's life span. The designation independent/dependent is clear in this case, because if a correlation is found, it cannot be that life span has influenced vitamin C intake, but an influence in the other direction is possible. Use in mathematics

Chapter 10. Comparing Two Groups

In calculus, a function is a map whose action is specified on variables. Take x and y to be two variables. A function f may map x to some expression in x. Assigning $y = f(x)$ gives a relation between x and y. If there is some relation specifying y in terms of x, then y is known as a 'dependent variable' (and x is an 'independent variable'). Use in statistics Controlled experiments

In a statistics experiment, the dependent variable is the event studied and expected to change whenever the independent variable is altered.

In the design of experiments, an independent variable's values are controlled or selected by the experimenter to determine its relationship to an observed phenomenon (i.e., the dependent variable). In such an experiment, an attempt is made to find evidence that the values of the independent variable determine the values of the dependent variable. The independent variable can be changed as required, and its values do not represent a problem requiring explanation in an analysis, but are taken simply as given. The dependent variable, on the other hand, usually cannot be directly controlled.

Controlled variables are also important to identify in experiments. They are the variables that are kept constant to prevent their influence on the effect of the independent variable on the dependent. Every experiment has a controlling variable, and it is necessary to not change it, or the results of the experiment won't be valid.

'Extraneous variables' are those that might affect the relationship between the independent and dependent variables. Extraneous variables are usually not theoretically interesting. They are measured in order for the experimenter to compensate for them. For example, an experimenter who wishes to measure the degree to which caffeine intake (the independent variable) influences explicit recall for a word list (the dependent variable) might also measure the participant's age (extraneous variable). She can then use these age data to control for the uninteresting effect of age, clarifying the relationship between caffeine and memory.

In summary:•Independent variables answer the question 'What do I change?'•Dependent variables answer the question 'What do I observe?'•Controlled variables answer the question 'What do I keep the same?'•Extraneous variables answer the question 'What uninteresting variables might mediate the effect of the IV on the DV?'Alternative terminology in statistics

In statistics, the dependent/independent variable terminology is used more widely than just in relation to controlled experiments. For example the data analysis of two jointly varying quantities may involve treating each in turn as the dependent variable and the other as the independent variable. However, for general usage, the pair _____ and explanatory variable is preferable as quantities treated as 'independent variables' are rarely statistically independent.

a. Ridge regression
b. Robust measures of scale
c. Response variable
d. Scale parameter

4. . _____ can be interpreted in mathematics, logic, linguistics, environmental science and other disciplines.

In its common meaning, the term is used to identify a characteristic, a feature, a measurable factor that can help in defining a particular system. It is an important element to take into consideration for the evaluation or for the comprehension of an event, a project or any situation.

Chapter 10. Comparing Two Groups

a. Parameter

b. Pathological

c. Pivotal quantity

d. Plugging in

5. _____ is the act of drawing a conclusion by deductive reasoning from given facts. The conclusion drawn is also called an _____. The laws of valid _____ are studied in the field of logic.

a. Inference

b. Analytic reasoning

c. Automated reasoning

d. Adaptive reasoning

1. c
2. d
3. c
4. a
5. a

You can take the complete Chapter Practice Test

for Chapter 10. Comparing Two Groups
on all key terms, persons, places, and concepts.

Online 99 Cents

http://www.epub27.14.20551.10.cram101.com/

Use www.Cram101.com for all your study needs

including Cram101's online interactive problem solving labs in

chemistry, statistics, mathematics, and more.

CHAPTER OUTLINE: KEY TERMS, PEOPLE, PLACES, CONCEPTS

Categorical variable

Statistic

Conditional distribution

Independence

Chi-squared distribution

Independent variable

Belief

Discrete probability distributions

Event

Cell

Chi-squared test

Test statistic

ANOVA

Degree

Freedom

Binomial

Contingency table

Goodness of fit

Ratio

	Relative risk
	Accident
	Residual analysis
	Standardized
	Ronald Aylmer Fisher
	Exact test

CHAPTER HIGHLIGHTS & NOTES: KEY TERMS, PEOPLE, PLACES, CONCEPTS

Categorical variable	Categorical Variables
	In statistics, a categorical variable is a variable that can take on one of a limited, and usually fixed, number of possible values. Categorical variables are often used to represent categorical data.
	A categorical variable that can take on exactly two values is termed a binary variable and is typically treated on its own as a special case.
Statistic	A statistic is a single measure of some attribute of a sample (e.g. its arithmetic mean value). It is calculated by applying a function (statistical algorithm) to the values of the items comprising the sample which are known together as a set of data.
	More formally, statistical theory defines a statistic as a function of a sample where the function itself is independent of the sample's distribution; that is, the function can be stated before realisation of the data.
Conditional distribution	Given two jointly distributed random variables X and Y, the conditional probability distribution of Y given X (written 'Y X') is the probability distribution of Y when X is known to be a particular value.

For discrete random variables, the conditional probability mass function can be written as P(Y = y X = x). From the definition of conditional probability, this is

X(y x) and this is

$$p_{Y|X}(y \mid x) = \frac{p_{X,Y}(x, y)}{p_X(x)} = \frac{p_{X|Y}(x \mid y)p_Y(y)}{p_X(x)},$$

where $p_{X,Y}$(x, y) gives the joint distribution of X and Y, while p_X(x) gives the marginal distribution for X.

The concept of the Conditional distribution of a continuous random variable is not as intuitive as it might seem: Borel's paradox shows that conditional probability density functions need not be invariant under coordinate transformations.

Independence	In probability theory, to say that two events are independent intuitively means that the occurrence of one event makes it neither more nor less probable that the other occurs. For example:•The event of getting a 6 the first time a die is rolled and the event of getting a 6 the second time are independent.•By contrast, the event of getting a 6 the first time a die is rolled and the event that the sum of the numbers seen on the first and second trials is 8 are not independent.•If two cards are drawn with replacement from a deck of cards, the event of drawing a red card on the first trial and that of drawing a red card on the second trial are independent.•By contrast, if two cards are drawn without replacement from a deck of cards, the event of drawing a red card on the first trial and that of drawing a red card on the second trial are again not independent. Similarly, two random variables are independent if the conditional probability distribution of either given the observed value of the other Is the same as If the other's value had not been observed. The concept of independence extends to dealing with collections of more than two events or random variables.
Chi-squared distribution	In probability theory and statistics, the chi-squared distribution with k degrees of freedom is the distribution of a sum of the squares of k independent standard normal random variables. It is one of the most widely used probability distributions in inferential statistics, e.g., in hypothesis testing or in construction of confidence intervals. When there is a need to contrast it with the noncentral chi-squared distribution, this distribution is sometimes called the central chi-squared distribution.
Independent variable	The terms 'dependent variable' and 'Independent variable' are used in similar but subtly different ways in mathematics and statistics as part of the standard terminology in those subjects.

They are used to distinguish between two types of quantities being considered, separating them into those available at the start of a process and those being created by it, where the latter (dependent variables) are dependent on the former (Independent variables).

The Independent variable is typically the variable being manipulated or changed and the dependent variable is the observed result of the Independent variable being manipulated.

Belief	Belief is the psychological state in which an individual holds a proposition or premise to be true.
	The terms Belief and knowledge are used differently in philosophy.
	Epistemology is the philosophical study of knowledge and Belief.
Discrete probability distributions	Discrete probability distributions arise in the mathematical description of probabilistic and statistical problems in which the values that might be observed are restricted to being within a pre-defined list of possible values. This list has either a finite number of members, or at most is countable.
	In probability theory, a probability distribution is called discrete if it is characterized by a probability mass function.
Event	In probability theory, an event is a set of outcomes (a subset of the sample space) to which a probability is assigned. Typically, when the sample space is finite, any subset of the sample space is an event (i.e. all elements of the power set of the sample space are defined as events). However, this approach does not work well in cases where the sample space is uncountably infinite, most notably when the outcome is a real number.
Cell	In geometry, a cell is a three-dimensional element that is part of a higher-dimensional object.
	In polytopes
	A cell is a three-dimensional polyhedron element that is part of the boundary of a higher-dimensional polytope, such as a polychoron (4-polytope) or honeycomb (3-space tessellation).
	For example, a cubic honeycomb is made of cubic cells, with 4 cubes on each edge.
Chi-squared test	A chi-squared test, also referred to as chi-square test or

χ^2 test, is any statistical hypothesis test in which the sampling distribution of the test statistic is a chi-squared distribution when the null hypothesis is true, or any in which this is asymptotically true, meaning that the sampling distribution (if the null hypothesis is true) can be made to approximate a chi-squared distribution as closely as desired by making the sample size large enough.

Some examples of chi-squared tests where the chi-squared distribution is only approximately valid:•Pearson's chi-squared test, also known as the chi-squared goodness-of-fit test or chi-squared test for independence. When mentioned without any modifiers or without other precluding context, this test is usually understood .•Yates's correction for continuity, also known as Yates' chi-squared test.•Cochran-Mantel-Haenszel chi-squared test.•McNemar's test, used in certain 2 × 2 tables with pairing•Linear-by-linear association chi-squared test•The portmanteau test in time-series analysis, testing for the presence of autocorrelation•Likelihood-ratio tests in general statistical modelling, for testing whether there is evidence of the need to move from a simple model to a more complicated one (where the simple model is nested within the complicated one).

One case where the distribution of the test statistic is an exact chi-squared distribution is the test that the variance of a normally distributed population has a given value based on a sample variance.

Test statistic	In statistical hypothesis testing, a hypothesis test is typically specified in terms of a test statistic, which is a function of the sample; it is considered as a numerical summary of a set of data that reduces the data to one or a small number of values that can be used to perform a hypothesis test. Given a null hypothesis and a test statistic T, we can specify a 'null value' T_0 such that values of T close to T_0 present the strongest evidence in favor of the null hypothesis, whereas values of T far from T_0 present the strongest evidence against the null hypothesis. An important property of a test statistic is that we must be able to determine its sampling distribution under the null hypothesis, which allows us to calculate p-values.
ANOVA	In statistics, ANOVA is a collection of statistical models, and their associated procedures, in which the observed variance is partitioned into components due to different sources of variation. In its simplest form ANOVA provides a statistical test of whether or not the means of several groups are all equal, and therefore generalizes Student's two-sample t-test to more than two groups. ANOVAs are helpful because they possess a certain advantage over a two-sample t-test. Doing multiple two-sample t-tests would result in a largely increased chance of committing a type I error. For this reason, ANOVAs are useful in comparing three or more means.

There are three conceptual classes of such models:

· Fixed-effects models assume that the data came from normal populations which may differ only in their means. (Model 1) · Random effects models assume that the data describe a hierarchy of different populations whose differences are constrained by the hierarchy. (Model 2) · Mixed-effect models describe the situations where both fixed and random effects are present. (Model 3)

Degree

In mathematics, there are several meanings of degree depending on the subject.

A degree (in full, a degree of arc, arc degree, or arcdegree), usually denoted by ° (the degree symbol), is a measurement of a plane angle, representing $\frac{1}{360}$ of a turn. When that angle is with respect to a reference meridian, it indicates a location along a great circle of a sphere, such as Earth , Mars, or the celestial sphere.

Freedom

Freedom (often referred to as the Freedom app) is a software program designed to keep a computer user away from the Internet for up to eight hours at a time. It is described as a way to 'free you from distractions, allowing you time to write, analyze, code, or create.' The program was written by Fred Stutzman, a Ph.D student at the University of North Carolina at Chapel Hill. Freedom is donationware.

Binomial

In elementary algebra, a Binomial is a polynomial with two terms--the sum of two monomials--often bound by parenthesis or brackets when operated upon. It is the simplest kind of polynomial other than monomials.

· The Binomial $a^2 - b^2$ can be factored as the product of two other Binomials:

$a^2 - b^2 = (a + b)(a - b)$.

This is a special case of the more general formula:

$$a^{n+1} - b^{n+1} = (a - b) \sum_{k=0}^{n} a^k b^{n-k}$$.

· The product of a pair of linear Binomials (ax + b) and (cx + d) is:

$(ax + b)(cx + d) = acx^2 + axd + bcx + bd$.

· A Binomial raised to the nth power, represented as

$(a + b)^n$

can be expanded by means of the Binomial theorem or, equivalently, using Pascal's triangle. Taking a simple example, the perfect square Binomial $(p + q)^2$ can be found by squaring the first term, adding twice the product of the first and second terms and finally adding the square of the second term, to give $p^2 + 2pq + q^2$.

· A simple but interesting application of the cited Binomial formula is the '(m,n)-formula' for generating Pythagorean triples: for m < n, let $a = n^2 - m^2$, $b = 2mn$, $c = n^2 + m^2$, then $a^2 + b^2 = c^2$.

Contingency table	In statistics, a contingency table is a type of table in a matrix format that displays the (multivariate) frequency distribution of the variables. It is often used to record and analyze the relation between two or more categorical variables. The term contingency table was first used by Karl Pearson in 'On the Theory of Contingency and Its Relation to Association and Normal Correlation', part of the Drapers' Company Research Memoirs Biometric Series I published in 1904.
Goodness of fit	The goodness of fit of a statistical model describes how well it fits a set of observations. Measures of goodness of fit typically summarize the discrepancy between observed values and the values expected under the model in question. Such measures can be used in statistical hypothesis testing, e.g. to test for normality of residuals, to test whether two samples are drawn from identical distributions , or whether outcome frequencies follow a specified distribution .
Ratio	A ratio is an expression that compares quantities relative to each other. The most common examples involve two quantities, but any number of quantities can be compared. ratios are represented mathematically by separating each quantity with a colon - for example, the ratio 2:3, which is read as the ratio 'two to three'.
Relative risk	In statistics and mathematical epidemiology, relative risk is the risk of an event relative to exposure. Relative risk is a ratio of the probability of the event occurring in the exposed group versus a non-exposed group. $$RR = \frac{p_{\text{exposed}}}{p_{\text{non-exposed}}}$$

Chapter 11. Analyzing the Association Between Categorical Variables

Accident	An accident is an unforeseen and unplanned event or circumstance, often with lack of intention or necessity. It usually implies a generally negative outcome which may have been avoided or prevented had circumstances leading up to the accident been recognized, and acted upon, prior to its occurrence.
	Experts in the field of injury prevention avoid use of the term 'accident' to describe events that cause injury in an attempt to highlight the predictable and preventable nature of most injuries.
Residual analysis	Residual analysis is a useful class of techniques for the evaluation of a fitted model. Checking the underlying assumptions is important since most linear regression estimators required a correctly specified regression function and independent distributed errors to be consistent.
Standardized	In mathematical statistics, a random variable X is standardized using the theoretical (population) mean and standard deviation:

$$Z = \frac{X - \mu}{\sigma}$$

where $\mu = E(X)$ is the mean and σ = the standard deviation of the probability distribution of X.

If the random variable under consideration is the sample mean:

$$\bar{X} = \frac{1}{n} \sum_{i=1}^{n} X_i$$

then the standardized version is

$$Z = \frac{\bar{X} - \mu}{\sigma / \sqrt{n}}.$$

Ronald Aylmer Fisher	Sir Ronald Aylmer Fisher FRS (17 February 1890 - 29 July 1962) was an English statistician, evolutionary biologist, eugenicist and geneticist. He was described by Anders Hald as 'a genius who almost single-handedly created the foundations for modern statistical science,' and Richard Dawkins described him as 'the greatest of Darwin's successors'.

| Exact test | In statistics, an exact (significance) test is a test where all assumptions upon which the derivation of the distribution of the test statistic is based are met, as opposed to an approximate test, in which the approximation may be made as close as desired by making the sample size big enough. This will result in a significance test that will have a false rejection rate always equal to the significance level of the test. For example an exact test at significance level 5% will in the long run reject true null hypothesis exactly 5% of the time. |

1. The terms 'dependent variable' and '_____' are used in similar but subtly different ways in mathematics and statistics as part of the standard terminology in those subjects. They are used to distinguish between two types of quantities being considered, separating them into those available at the start of a process and those being created by it, where the latter (dependent variables) are dependent on the former (_____s).

   The _____ is typically the variable being manipulated or changed and the dependent variable is the observed result of the _____ being manipulated.

   a. Absolute value
   b. Independent variable
   c. affinely extended real number system
   d. Aircraft design

2. . Given two jointly distributed random variables X and Y, the conditional probability distribution of Y given X (written 'Y X') is the probability distribution of Y when X is known to be a particular value.

   For discrete random variables, the conditional probability mass function can be written as P(Y = y X = x). From the definition of conditional probability, this is

   X(y x) and this is

   $$p_{Y|X}(y \mid x) = \frac{p_{X,Y}(x, y)}{p_X(x)} = \frac{p_{X|Y}(x \mid y)p_Y(y)}{p_X(x)},$$

   where $p_{X,Y}(x, y)$ gives the joint distribution of X and Y, while $p_X(x)$ gives the marginal distribution for X.

   The concept of the _____ of a continuous random variable is not as intuitive as it might seem: Borel's paradox shows that conditional probability density functions need not be invariant under coordinate transformations.

   a. sample space

    b. Conditional distribution

    c. density function

    d. Probability theory

3. In probability theory and statistics, the _____ with k degrees of freedom is the distribution of a sum of the squares of k independent standard normal random variables. It is one of the most widely used probability distributions in inferential statistics, e.g., in hypothesis testing or in construction of confidence intervals. When there is a need to contrast it with the noncentral _____, this distribution is sometimes called the central _____.

    a. Circular uniform distribution

    b. Complex normal distribution

    c. Chi-squared distribution

    d. Davis distribution

4. _____s

    In statistics, a _____ is a variable that can take on one of a limited, and usually fixed, number of possible values. _____s are often used to represent categorical data.

    A _____ that can take on exactly two values is termed a binary variable and is typically treated on its own as a special case.

    a. Censoring

    b. Compositional data

    c. Categorical variable

    d. Cross-sectional data

5. A _____ is a single measure of some attribute of a sample (e.g. its arithmetic mean value). It is calculated by applying a function (statistical algorithm) to the values of the items comprising the sample which are known together as a set of data.

    More formally, statistical theory defines a _____ as a function of a sample where the function itself is independent of the sample's distribution; that is, the function can be stated before realisation of the data.

    a. Bayesian inference

    b. Berkson error model

    c. Statistic

    d. Conditionality principle

1. b
2. b
3. c
4. c
5. c

You can take the complete Chapter Practice Test

for Chapter 11. Analyzing the Association Between Categorical Variables
on all key terms, persons, places, and concepts.

Online 99 Cents

http://www.epub27.14.20551.11.cram101.com/

Use www.Cram101.com for all your study needs

including Cram101's online interactive problem solving labs in

chemistry, statistics, mathematics, and more.

Correlation

Regression analysis

Regression line

Maximum

Y-intercept

ANOVA

Least squares

Parameter

Regression equation

Conditional distribution

Chi-squared distribution

Internet

Paradox

Statistic

Squared

Placebo effect

Predictive power

Proportional reduction in error

Residual sum of squares

CHAPTER OUTLINE: KEY TERMS, PEOPLE, PLACES, CONCEPTS

Total sum of squares

Single event upset

Factorization

Two-way analysis

Variance

Ecological fallacy

Fallacies

Statistical inference

Independence

Inference

Confidence interval

Estimation

Standardized

Histogram

Deviation

Mean square error

Residual standard deviation

Standard deviation

Prediction interval

	Exponential growth

Correlation	In statistics, correlation (often measured as a correlation coefficient, ρ) indicates the strength and direction of a relationship between two random variables. The commonest use refers to a linear relationship. In general statistical usage, correlation or co-relation refers to the departure of two random variables from independence.
Regression analysis	In statistics, regression analysis includes many techniques for modeling and analyzing several variables, when the focus is on the relationship between a dependent variable and one or more independent variables. More specifically, regression analysis helps one understand how the typical value of the dependent variable changes when any one of the independent variables is varied, while the other independent variables are held fixed. Most commonly, regression analysis estimates the conditional expectation of the dependent variable given the independent variables -- that is, the average value of the dependent variable when the independent variables are held fixed.
Regression line	Regression line is a line drawn through a scatterplot of two variables. The line is chosen so that it comes as close to the points as possible.
Maximum	The largest and the smallest element of a set are called extreme values, absolute extrema, extreme records, or optima.

For a differentiable function f, if $f(x_0)$ is an extreme value for the set of all values f(x), and if x_0 is in the interior of the domain of f, then x_0 is a critical point, by Fermat's theorem.

The point or points at which a function assumes its maximum (respectively, minimum) value are called the arg max (respectively, arg min): the arguments (inputs) at which the maximum (respectively, minimum) occurs. |
| Y-intercept | In coordinate geometry, using the common convention that the horizontal axis represents a variable x and the vertical axis represents a variable y, a y-intercept is a point where the graph of a function or relation intersects with the y-axis of the coordinate system. As such, these points satisfy x=0. |

ANOVA	In statistics, ANOVA is a collection of statistical models, and their associated procedures, in which the observed variance is partitioned into components due to different sources of variation. In its simplest form ANOVA provides a statistical test of whether or not the means of several groups are all equal, and therefore generalizes Student's two-sample t-test to more than two groups. ANOVAs are helpful because they possess a certain advantage over a two-sample t-test. Doing multiple two-sample t-tests would result in a largely increased chance of committing a type I error. For this reason, ANOVAs are useful in comparing three or more means.
	There are three conceptual classes of such models:
	· Fixed-effects models assume that the data came from normal populations which may differ only in their means. (Model 1) · Random effects models assume that the data describe a hierarchy of different populations whose differences are constrained by the hierarchy. (Model 2) · Mixed-effect models describe the situations where both fixed and random effects are present. (Model 3)
Least squares	The method of least squares is a standard approach to the approximate solution of overdetermined systems, i.e., sets of equations in which there are more equations than unknowns. 'Least squares' means that the overall solution minimizes the sum of the squares of the errors made in the results of every single equation.
	The most important application is in data fitting.
Parameter	Parameter can be interpreted in mathematics, logic, linguistics, environmental science and other disciplines.
	In its common meaning, the term is used to identify a characteristic, a feature, a measurable factor that can help in defining a particular system. It is an important element to take into consideration for the evaluation or for the comprehension of an event, a project or any situation.
Regression equation	The regression equation represents the relation between selected values of one variable (x) and observed values of the other (y); it permits the prediction of the most probable values of y.
Conditional distribution	Given two jointly distributed random variables X and Y, the conditional probability distribution of Y given X (written 'Y X') is the probability distribution of Y when X is known to be a particular value.
	For discrete random variables, the conditional probability mass function can be written as $P(Y = y X = x)$. From the definition of conditional probability, this is
	X(y x) and this is

$$p_{Y|X}(y \mid x) = \frac{p_{X,Y}(x, y)}{p_X(x)} = \frac{p_{X|Y}(x \mid y)p_Y(y)}{p_X(x)},$$

where $p_{X,Y}(x, y)$ gives the joint distribution of X and Y, while $p_X(x)$ gives the marginal distribution for X.

The concept of the Conditional distribution of a continuous random variable is not as intuitive as it might seem: Borel's paradox shows that conditional probability density functions need not be invariant under coordinate transformations.

Chi-squared distribution	In probability theory and statistics, the chi-squared distribution with k degrees of freedom is the distribution of a sum of the squares of k independent standard normal random variables. It is one of the most widely used probability distributions in inferential statistics, e.g., in hypothesis testing or in construction of confidence intervals. When there is a need to contrast it with the noncentral chi-squared distribution, this distribution is sometimes called the central chi-squared distribution.
Internet	The Internet is a global system of interconnected computer networks that use the standard Internet Protocol Suite (TCP/IP) to serve billions of users worldwide. It is a network of networks that consists of millions of private, public, academic, business, and government networks, of local to global scope, that are linked by a broad array of electronic and optical networking technologies. The Internet carries a vast range of information resources and services, such as the inter-linked hypertext documents of the World Wide Web (WWW) and the infrastructure to support electronic mail.
Paradox	A Paradox is a true statement or group of statements that leads to a contradiction or a situation which defies intuition. The term is also used for an apparent contradiction that actually expresses a non-dual truth (cf. kÅan, Catuskoti).
Statistic	A statistic is a single measure of some attribute of a sample (e.g. its arithmetic mean value). It is calculated by applying a function (statistical algorithm) to the values of the items comprising the sample which are known together as a set of data.

More formally, statistical theory defines a statistic as a function of a sample where the function itself is independent of the sample's distribution; that is, the function can be stated before realisation of the data. |
| Squared | In algebra, the square of a number is that number multiplied by itself. To square a quantity is to multiply it by itself. Its notation is a superscripted '2'; a number x squared is written as x^2. |

Placebo effect	Placebo effect is the term applied by medical science to the therapeutical and healing effects of inert medicines and/or ritualistic or faith healing practices. The placebo effect occurs when a patient takes an inert substance . Experimenters typically use placebos in the context of a clinical trial, in which a 'test group' of patients receives the therapy being tested, and a 'control group' receives the placebo. It can then be determined if results from the 'test' group exceed those due to the placebo effect.
Predictive power	The predictive power of a scientific theory refers to its ability to generate testable predictions. Theories with strong predictive power are highly valued, because the predictions can often encourage the falsification of the theory. The concept of predictive power differs from explanatory and descriptive power (where phenomena that are already known are retrospectively explained by a given theory) in that it allows a prospective test of theoretical understanding.
Proportional reduction in error	Proportional reduction in loss (PRL) refers to a general framework for developing and evaluating measures of the reliability of particular ways of making observations which are possibly subject to errors of all types. Such measures quantify how much having the observations available has reduced the loss (cost) of the uncertainty about the intended quantity compared with not having those observations. proportional reduction in error is a more restrictive framework widely used in statistics, in which the general loss function is replaced by a more direct measure of error such as the mean square error.
Residual sum of squares	In statistics, the residual sum of squares is the sum of squares of residuals. It is also known as the sum of squared residuals (SSR) or the sum of squared errors of prediction (SSE). It is a measure of the discrepancy between the data and an estimation model.
Total sum of squares	In statistical data analysis the total sum of squares is a quantity that appears as part of a standard way of presenting results of such analyses. It is defined as being the sum, over all observations, of the squared differences of each observation from the overall mean. In statistical linear models, (particularly in standard regression models), the TSS is the sum of the squares of the difference of the dependent variable and its grand mean: $\sum_{i=1}^{n}(y_i - \bar{y})^2$. For wide classes of linear models: Total sum of squares = explained sum of squares + residual sum of squares.
Single event upset	A single event upset is a change of state caused by ions or electro-magnetic radiation striking a sensitive node in a micro-electronic device, such as in a microprocessor, semiconductor memory).

Factorization	In mathematics, Factorization or factoring is the decomposition of an object into a product of other objects, which when multiplied together give the original. For example, the number 15 factors into primes as 3 × 5, and the polynomial $x^2 - 4$ factors as $(x - 2)(x + 2)$. In all cases, a product of simpler objects is obtained.
Two-way analysis	Two-way analysis is the computation of two variances price and quantity variances for direct materials and direct labor and budget and volume variances for factory overhead.
Variance	In probability theory and statistics, the variance is a measure of how far a set of numbers is spread out. It is one of several descriptors of a probability distribution, describing how far the numbers lie from the mean (expected value). In particular, the variance is one of the moments of a distribution.
Ecological fallacy	An ecological fallacy is a logical fallacy in the interpretation of statistical data in an ecological study, whereby inferences about the nature of individuals are based solely upon aggregate statistics collected for the group to which those individuals belong. In epidemiology, the ecological fallacy is committed when a correlation observed at the population level is assumed to apply at the individual level. This fallacy assumes that individual members of a group have the average characteristics of the group at large.
Fallacies	In logic and rhetoric, a fallacy is a misconception resulting from incorrect reasoning in argumentation. By accident or design, fallacies may exploit emotional triggers in the listener or interlocutor (e.g. appeal to emotion), or take advantage of social relationships between people . Fallacious arguments are often structured using rhetorical patterns that obscure the logical argument, making fallacies more difficult to diagnose.
Statistical inference	In statistics, statistical inference is the process of drawing conclusions from data subject to random variation, for example, observational errors or sampling variation. More substantially, the terms statistical inference, statistical induction and inferential statistics are used to describe systems of procedures that can be used to draw conclusions from datasets arising from systems affected by random variation, such as observational errors, random sampling, or random experimentation. Initial requirements of such a system of procedures for inference and induction are that the system should produce reasonable answers when applied to well-defined situations and that it should be general enough to be applied across a range of situations.
Independence	In probability theory, to say that two events are independent intuitively means that the occurrence of one event makes it neither more nor less probable that the other occurs.

	For example:•The event of getting a 6 the first time a die is rolled and the event of getting a 6 the second time are independent.•By contrast, the event of getting a 6 the first time a die is rolled and the event that the sum of the numbers seen on the first and second trials is 8 are not independent.•If two cards are drawn with replacement from a deck of cards, the event of drawing a red card on the first trial and that of drawing a red card on the second trial are independent.•By contrast, if two cards are drawn without replacement from a deck of cards, the event of drawing a red card on the first trial and that of drawing a red card on the second trial are again not independent.
	Similarly, two random variables are independent if the conditional probability distribution of either given the observed value of the other is the same as if the other's value had not been observed. The concept of independence extends to dealing with collections of more than two events or random variables.
Inference	Inference is the act of drawing a conclusion by deductive reasoning from given facts. The conclusion drawn is also called an inference. The laws of valid inference are studied in the field of logic.
Confidence interval	In statistics, a confidence interval is a kind of interval estimate of a population parameter and is used to indicate the reliability of an estimate. It is an observed interval (i.e. it is calculated from the observations), in principle different from sample to sample, that frequently includes the parameter of interest, if the experiment is repeated. How frequently the observed interval contains the parameter is determined by the confidence level or confidence coefficient.
Estimation	In project management (i.e., for engineering), accurate estimates are the basis of sound project planning. Many processes have been developed to aid engineers in making accurate estimates, such as•Analogy based estimation•Compartmentalization (i.e., breakdown of tasks)•Delphi method•Documenting estimation results•Educated assumptions•Estimating each task•Examining historical data•Identifying dependencies•Parametric estimating•Risk assessment•Structured planning
	Popular estimation processes for software projects include:•Cocomo•Cosysmo•Event chain methodology•Function points•Program Evaluation and Review Technique (PERT)•Proxy Based Estimation (PROBE) (from the Personal Software Process)•The Planning Game (from Extreme Programming)•Weighted Micro Function Points (WMFP)•Wideband Delphi.
Standardized	In mathematical statistics, a random variable X is standardized using the theoretical (population) mean and standard deviation:

$$Z = \frac{X - \mu}{\sigma}$$

where $\mu = E(X)$ is the mean and $\sigma =$ the standard deviation of the probability distribution of X.

If the random variable under consideration is the sample mean:

$$\bar{X} = \frac{1}{n} \sum_{i=1}^{n} X_i$$

then the standardized version is

$$Z = \frac{\bar{X} - \mu}{\sigma / \sqrt{n}}.$$

Histogram	In statistics, a histogram is a graphical representation showing a visual impression of the distribution of data. It is an estimate of the probability distribution of a continuous variable and was first introduced by Karl Pearson. A histogram consists of tabular frequencies, shown as adjacent rectangles, erected over discrete intervals (bins), with an area equal to the frequency of the observations in the interval.
Deviation	In mathematics and statistics, deviation is a measure of difference between the observed value and the mean. The sign of deviation (positive or negative), reports the direction of that difference (it is larger when the sign is positive, and smaller if it is negative). The magnitude of the value indicates the size of the difference.
Mean square error	In statistics, the mean square error is one of many ways to quantify the difference between values implied by an estimator and the true values of the quantity being estimated. Mean square error is a risk function, corresponding to the expected value of the squared error loss or quadratic loss. Mean square error measures the average of the squares of the 'errors.' The error is the amount by which the value implied by the estimator differs from the quantity to be estimated.
Residual standard deviation	The residual standard deviation is a goodness-fit measure. The smaller the residual standard deviation, the closer is the fit to the data.
Standard deviation	Standard deviation is a widely used measurement of variability or diversity used in statistics and probability theory.

It shows how much variation or 'dispersion' there is from the 'average' (mean, or expected/budgeted value). A low standard deviation indicates that the data points tend to be very close to the mean, whereas high standard deviation indicates that the data are spread out over a large range of values.

Prediction interval

In statistical inference, specifically predictive inference, a prediction interval is an estimate of an interval in which future observations will fall, with a certain probability, given what has already been observed. Prediction intervals are often used in regression analysis.

Prediction intervals are used in both frequentist statistics and Bayesian statistics: a prediction interval bears the same relationship to a future observation that a frequentist confidence interval or Bayesian credible interval bears to an unobservable population parameter: prediction intervals predict the distribution of individual future points, whereas confidence intervals and credible intervals of parameters predict the distribution of estimates of the true population mean or other quantity of interest that cannot be observed.

Exponential growth

Exponential growth occurs when the growth rate of the value of a mathematical function is proportional to the function's current value. In the case of a discrete domain of definition with equal intervals it is also called geometric growth or geometric decay (the function values form a geometric progression).

The formula for exponential growth of a variable x at the (positive or negative) growth rate r, as time t goes on in discrete intervals (that is, at integer times 0, 1, 2, 3, .). is $x_t = x_0(1 + r)^t$

where x_0 is the value of x at time 0. For example, with a growth rate of r = 5% = 0.05, going from any integer value of time to the next integer causes x at the second time to be 1.05 times (i.e., 5% larger than) what it was at the previous time.

1. In logic and rhetoric, a fallacy is a misconception resulting from incorrect reasoning in argumentation. By accident or design, _____ may exploit emotional triggers in the listener or interlocutor (e.g. appeal to emotion), or take advantage of social relationships between people . Fallacious arguments are often structured using rhetorical patterns that obscure the logical argument, making _____ more difficult to diagnose.

   a. Fallacy of four terms
   b. Propositional calculus
   c. 1-factor
   d. Fallacies

2. _____ can be interpreted in mathematics, logic, linguistics, environmental science and other disciplines.

   In its common meaning, the term is used to identify a characteristic, a feature, a measurable factor that can help in defining a particular system. It is an important element to take into consideration for the evaluation or for the comprehension of an event, a project or any situation.

   a. Parts-per notation
   b. Parameter
   c. Pivotal quantity
   d. Plugging in

3. In statistics, the _____ is one of many ways to quantify the difference between values implied by an estimator and the true values of the quantity being estimated. _____ is a risk function, corresponding to the expected value of the squared error loss or quadratic loss. _____ measures the average of the squares of the 'errors.' The error is the amount by which the value implied by the estimator differs from the quantity to be estimated.

   a. Coefficient of determination
   b. Mean square error
   c. Partial least squares regression
   d. Mean squared error

4. _____ is a line drawn through a scatterplot of two variables. The line is chosen so that it comes as close to the points as possible.

   a. Composite measure
   b. Relative-frequency distribution
   c. Regression line
   d. Frequency histogram

5. . A _____ is a true statement or group of statements that leads to a contradiction or a situation which defies intuition. The term is also used for an apparent contradiction that actually expresses a non-dual truth (cf. kÅan, Catuskoti).

   a. 1-factor
   b. Continuant
   c. Generalized inverse

1. d
2. b
3. b
4. c
5. d

You can take the complete Chapter Practice Test

for Chapter 12. Analyzing the Association Between Quantitative Variables: Regression Analysis
on all key terms, persons, places, and concepts.

Online 99 Cents

http://www.epub27.14.20551.12.cram101.com/

Use www.Cram101.com for all your study needs

including Cram101's online interactive problem solving labs in

chemistry, statistics, mathematics, and more.

Chapter 13. Multiple Regression

CHAPTER OUTLINE: KEY TERMS, PEOPLE, PLACES, CONCEPTS

_____	Regression equation
_____	Multiple correlation
_____	Squared
_____	Correlation
_____	Paradox
_____	ANOVA
_____	Mean square error
_____	Statistical inference
_____	Deviation
_____	Inference
_____	Standard deviation
_____	Parameter
_____	Statistic
_____	Test statistic
_____	Variance
_____	Confidence interval
_____	Backward elimination
_____	Forward selection
_____	Stepwise regression

_____ | Building _____

_____ | Logistic regression

_____ | Relative standing

_____ | Model checking

CHAPTER HIGHLIGHTS & NOTES: KEY TERMS, PEOPLE, PLACES, CONCEPTS

Regression equation	The regression equation represents the relation between selected values of one variable (x) and observed values of the other (y); it permits the prediction of the most probable values of y.
Multiple correlation	In statistics, multiple correlation is a linear relationship among more than two variables. It is measured by the coefficient of multiple determination, denoted as R^2, which is a measure of the fit of a linear regression. A regression's R^2 falls somewhere between zero and one (assuming a constant term has been included in the regression); a higher value indicates a stronger relationship among the variables, with a value of one indicating that all data points fall exactly on a line in multidimensional space and a value of zero indicating no relationship at all between the independent variables collectively and the dependent variable.
Squared	In algebra, the square of a number is that number multiplied by itself. To square a quantity is to multiply it by itself. Its notation is a superscripted '2'; a number x squared is written as x^2.
Correlation	In statistics, correlation (often measured as a correlation coefficient, ρ) indicates the strength and direction of a relationship between two random variables. The commonest use refers to a linear relationship. In general statistical usage, correlation or co-relation refers to the departure of two random variables from independence.
Paradox	A Paradox is a true statement or group of statements that leads to a contradiction or a situation which defies intuition. The term is also used for an apparent contradiction that actually expresses a non-dual truth (cf. kÅan, Catuskoti).

ANOVA	In statistics, ANOVA is a collection of statistical models, and their associated procedures, in which the observed variance is partitioned into components due to different sources of variation. In its simplest form ANOVA provides a statistical test of whether or not the means of several groups are all equal, and therefore generalizes Student's two-sample t-test to more than two groups. ANOVAs are helpful because they possess a certain advantage over a two-sample t-test. Doing multiple two-sample t-tests would result in a largely increased chance of committing a type I error. For this reason, ANOVAs are useful in comparing three or more means.

There are three conceptual classes of such models:

· Fixed-effects models assume that the data came from normal populations which may differ only in their means. (Model 1) · Random effects models assume that the data describe a hierarchy of different populations whose differences are constrained by the hierarchy. (Model 2) · Mixed-effect models describe the situations where both fixed and random effects are present. (Model 3) |
Mean square error	In statistics, the mean square error is one of many ways to quantify the difference between values implied by an estimator and the true values of the quantity being estimated. Mean square error is a risk function, corresponding to the expected value of the squared error loss or quadratic loss. Mean square error measures the average of the squares of the 'errors.' The error is the amount by which the value implied by the estimator differs from the quantity to be estimated.
Statistical inference	In statistics, statistical inference is the process of drawing conclusions from data subject to random variation, for example, observational errors or sampling variation. More substantially, the terms statistical inference, statistical induction and inferential statistics are used to describe systems of procedures that can be used to draw conclusions from datasets arising from systems affected by random variation, such as observational errors, random sampling, or random experimentation. Initial requirements of such a system of procedures for inference and induction are that the system should produce reasonable answers when applied to well-defined situations and that it should be general enough to be applied across a range of situations.
Deviation	In mathematics and statistics, deviation is a measure of difference between the observed value and the mean. The sign of deviation (positive or negative), reports the direction of that difference (it is larger when the sign is positive, and smaller if it is negative). The magnitude of the value indicates the size of the difference.
Inference	Inference is the act of drawing a conclusion by deductive reasoning from given facts. The conclusion drawn is also called an inference. The laws of valid inference are studied in the field of logic.

Chapter 13. Multiple Regression

Standard deviation	Standard deviation is a widely used measurement of variability or diversity used in statistics and probability theory. It shows how much variation or 'dispersion' there is from the 'average' (mean, or expected/budgeted value). A low standard deviation indicates that the data points tend to be very close to the mean, whereas high standard deviation indicates that the data are spread out over a large range of values.
Parameter	Parameter can be interpreted in mathematics, logic, linguistics, environmental science and other disciplines. In its common meaning, the term is used to identify a characteristic, a feature, a measurable factor that can help in defining a particular system. It is an important element to take into consideration for the evaluation or for the comprehension of an event, a project or any situation.
Statistic	A statistic is a single measure of some attribute of a sample (e.g. its arithmetic mean value). It is calculated by applying a function (statistical algorithm) to the values of the items comprising the sample which are known together as a set of data. More formally, statistical theory defines a statistic as a function of a sample where the function itself is independent of the sample's distribution; that is, the function can be stated before realisation of the data.
Test statistic	In statistical hypothesis testing, a hypothesis test is typically specified in terms of a test statistic, which is a function of the sample; it is considered as a numerical summary of a set of data that reduces the data to one or a small number of values that can be used to perform a hypothesis test. Given a null hypothesis and a test statistic T, we can specify a 'null value' T_0 such that values of T close to T_0 present the strongest evidence in favor of the null hypothesis, whereas values of T far from T_0 present the strongest evidence against the null hypothesis. An important property of a test statistic is that we must be able to determine its sampling distribution under the null hypothesis, which allows us to calculate p-values.
Variance	In probability theory and statistics, the variance is a measure of how far a set of numbers is spread out. It is one of several descriptors of a probability distribution, describing how far the numbers lie from the mean (expected value). In particular, the variance is one of the moments of a distribution.
Confidence interval	In statistics, a confidence interval is a kind of interval estimate of a population parameter and is used to indicate the reliability of an estimate. It is an observed interval (i.e. it is calculated from the observations), in principle different from sample to sample, that frequently includes the parameter of interest, if the experiment is repeated. How frequently the observed interval contains the parameter is determined by the confidence level or confidence coefficient.

Backward elimination	In statistics, stepwise regression includes regression models in which the choice of predictive variables is carried out by an automatic procedure. Usually, this takes the form of a sequence of F-tests, but other techniques are possible, such as t-tests, adjusted R-square, Akaike information criterion, Bayesian information criterion, Mallows' Cp, or false discovery rate.

The main approaches are:

· Forward selection, which involves starting with no variables in the model, trying out the variables one by one and including them if they are 'statistically significant'. · Backward elimination, which involves starting with all candidate variables and testing them one by one for statistical significance, deleting any that are not significant. · Methods that are a combination of the above, testing at each stage for variables to be included or excluded.

A widely used algorithm was first proposed by Efroymson (1960).

Forward selection	In statistics, stepwise regression includes regression models in which the choice of predictive variables is carried out by an automatic procedure. Usually, this takes the form of a sequence of F-tests, but other techniques are possible, such as t-tests, adjusted R-square, Akaike information criterion, Bayesian information criterion, Mallows' Cp, or false discovery rate.

The main approaches are:

· Forward selection, which involves starting with no variables in the model, trying out the variables one by one and including them if they are 'statistically significant'. · Backward elimination, which involves starting with all candidate variables and testing them one by one for statistical significance, deleting any that are not significant. · Methods that are a combination of the above, testing at each stage for variables to be included or excluded.

A widely used algorithm was first proposed by Efroymson (1960).

Stepwise regression	In statistics, stepwise regression includes regression models in which the choice of predictive variables is carried out by an automatic procedure. Usually, this takes the form of a sequence of F-tests, but other techniques are possible, such as t-tests, adjusted R-square, Akaike information criterion, Bayesian information criterion, Mallows' Cp, or false discovery rate.

The main approaches are:•Forward selection, which involves starting with no variables in the model, trying out the variables one by one and including them if they are 'statistically significant'.•Backward elimination, which involves starting with all candidate variables and testing them one by one for statistical significance, deleting any that are not significant.•Methods that are a combination of the above, testing at each stage for variables to be included or excluded.

Chapter 13. Multiple Regression

Building	In mathematics, a building (also Tits building, Bruhat-Tits building, finite projective planes, and Riemannian symmetric spaces. Initially introduced by Jacques Tits as a means to understand the structure of exceptional groups of Lie type, the theory has also been used to study the geometry and topology of homogeneous spaces of p-adic Lie groups and their discrete subgroups of symmetries, in the same way that trees have been used to study free groups. The notion of a building was invented by Jacques Tits as a means of describing simple algebraic groups over an arbitrary field.
Logistic regression	In statistics, logistic regression is a type of regression analysis used for predicting the outcome of a categorical (a variable that can take on a limited number of categories) criterion variable based on one or more predictor variables. Logistic regression can be bi- or multinomial. Binomial or binary logistic regression refers to the instance in which the criterion can take on only two possible outcomes (e.g., 'dead' vs. 'alive', 'success' vs. 'failure', or 'yes' vs. 'no').
Relative standing	Relative standing is a measurement of numbers which indicate where a particular values lies in relation to the rest of the values in a set of data or population.
Model checking	In the field of logic in computer science, model checking refers to the following problem: Given a model of a system, test automatically whether this model meets a given specification. Typically, the systems one has in mind are hardware or software systems, and the specification contains safety requirements such as the absence of deadlocks and similar critical states that can cause the system to crash. Model checking is a technique for automatically verifying correctness properties of finite-state systems.

Chapter 13. Multiple Regression

1. The _____ represents the relation between selected values of one variable (x) and observed values of the other (y); it permits the prediction of the most probable values of y.

   a. 1-factor
   b. Regression equation
   c. Vector
   d. BioSense

2. In probability theory and statistics, the _____ is a measure of how far a set of numbers is spread out. It is one of several descriptors of a probability distribution, describing how far the numbers lie from the mean (expected value). In particular, the _____ is one of the moments of a distribution.

   a. Variation ratio
   b. Variogram
   c. Deviance
   d. Variance

3. In statistics, _____ is a type of regression analysis used for predicting the outcome of a categorical (a variable that can take on a limited number of categories) criterion variable based on one or more predictor variables. _____ can be bi- or multinomial. Binomial or binary _____ refers to the instance in which the criterion can take on only two possible outcomes (e.g., 'dead' vs. 'alive', 'success' vs. 'failure', or 'yes' vs. 'no').

   a. Logit
   b. Multinomial test
   c. Logistic regression
   d. Nominal category

4. In statistics, _____ is a linear relationship among more than two variables. It is measured by the coefficient of multiple determination, denoted as R^2, which is a measure of the fit of a linear regression. A regression's R^2 falls somewhere between zero and one (assuming a constant term has been included in the regression); a higher value indicates a stronger relationship among the variables, with a value of one indicating that all data points fall exactly on a line in multidimensional space and a value of zero indicating no relationship at all between the independent variables collectively and the dependent variable.

   a. Nonparametric regression
   b. Multiple correlation
   c. Path analysis
   d. Path coefficient

5. . A _____ is a true statement or group of statements that leads to a contradiction or a situation which defies intuition. The term is also used for an apparent contradiction that actually expresses a non-dual truth (cf. kÅan, Catuskoti).

   a. Paradox
   b. Pearson product-moment correlation coefficient
   c. Sample covariance

1. b
2. d
3. c
4. b
5. a

You can take the complete Chapter Practice Test

for Chapter 13. Multiple Regression
on all key terms, persons, places, and concepts.

Online 99 Cents

http://www.epub27.14.20551.13.cram101.com/

Use www.Cram101.com for all your study needs

including Cram101's online interactive problem solving labs in

chemistry, statistics, mathematics, and more.

CHAPTER OUTLINE: KEY TERMS, PEOPLE, PLACES, CONCEPTS

_____ | Variance

_____ | ANOVA

_____ | Factorization

_____ | One-way ANOVA

_____ | Factorial ANOVA

_____ | Statistic

_____ | Test statistic

_____ | Total sum of squares

_____ | T-test

_____ | Confidence interval

_____ | Holm-Bonferroni method

_____ | Carlo Emilio Bonferroni

_____ | Exploratory data analysis

_____ | Multiple comparisons

_____ | Tukey method

_____ | Data analysis

_____ | Regression analysis

_____ | Ronald Aylmer Fisher

_____ | Mains

_____ | Main effect

_____ | Inference

_____ | Factorial

_____ | Ranking

_____ | Financial statement

_____ | P-value

_____ | Reaction time

_____ | Estimation

_____ | Median

_____ | Interquartile range

_____ | Sign test

_____ | Internet

_____ | Browsing

_____ | Binomial

_____ | Binomial distribution

_____ | Wilcoxon signed-ranks test

_____ | Chi-squared test

_____ | Null hypothesis

_____ | Correlation

	Independence
	Multiple correlation
	Standardized
	Logistic regression
	Sampling distribution

CHAPTER HIGHLIGHTS & NOTES: KEY TERMS, PEOPLE, PLACES, CONCEPTS

Variance	In probability theory and statistics, the variance is a measure of how far a set of numbers is spread out. It is one of several descriptors of a probability distribution, describing how far the numbers lie from the mean (expected value). In particular, the variance is one of the moments of a distribution.
ANOVA	In statistics, ANOVA is a collection of statistical models, and their associated procedures, in which the observed variance is partitioned into components due to different sources of variation. In its simplest form ANOVA provides a statistical test of whether or not the means of several groups are all equal, and therefore generalizes Student's two-sample t-test to more than two groups. ANOVAs are helpful because they possess a certain advantage over a two-sample t-test. Doing multiple two-sample t-tests would result in a largely increased chance of committing a type I error. For this reason, ANOVAs are useful in comparing three or more means.

There are three conceptual classes of such models:

· Fixed-effects models assume that the data came from normal populations which may differ only in their means. (Model 1) · Random effects models assume that the data describe a hierarchy of different populations whose differences are constrained by the hierarchy. (Model 2) · Mixed-effect models describe the situations where both fixed and random effects are present. (Model 3)

Chapter 14. Comparing Groups: Analysis of Variance Methods

Factorization	In mathematics, Factorization or factoring is the decomposition of an object into a product of other objects, which when multiplied together give the original. For example, the number 15 factors into primes as 3 × 5, and the polynomial $x^2 - 4$ factors as $(x - 2)(x + 2)$. In all cases, a product of simpler objects is obtained.
One-way ANOVA	In statistics, One-way ANOVA is a technique used to compare means of two or more samples (using the F distribution). This technique can be used only for numerical data.
	The ANOVA tests the null hypothesis that samples in two or more groups are drawn from the same population. To do this, two estimates are made of the population variance. These estimates rely on various assumptions. The ANOVA produces an F statistic, the ratio of the variance calculated among the means to the variance within the samples. If the group means are drawn from the same population, the variance between the group means should be lower than the variance of the samples, following central limit theorem. A higher ratio therefore implies that the samples were drawn from different populations.
Factorial ANOVA	Factorial ANOVA is used when the experimenter wants to study the effects of two or more treatment variables. The most commonly used type of Factorial ANOVA is the 2^2 (read 'two by two') design, where there are two independent variables and each variable has two levels or distinct values. However, such use of ANOVA for analysis of 2^k factorial designs and fractional factorial designs is 'confusing and makes little sense'; instead it is suggested to refer the value of the effect divided by its standard error to a t-table. Factorial ANOVA can also be multi-level such as 3^3, etc. or higher order such as 2×2×2, etc.. Since the introduction of data analytic software, the utilization of higher order designs and analyses has become quite common.
Statistic	A statistic is a single measure of some attribute of a sample (e.g. its arithmetic mean value). It is calculated by applying a function (statistical algorithm) to the values of the items comprising the sample which are known together as a set of data.
	More formally, statistical theory defines a statistic as a function of a sample where the function itself is independent of the sample's distribution; that is, the function can be stated before realisation of the data.
Test statistic	In statistical hypothesis testing, a hypothesis test is typically specified in terms of a test statistic, which is a function of the sample; it is considered as a numerical summary of a set of data that reduces the data to one or a small number of values that can be used to perform a hypothesis test. Given a null hypothesis and a test statistic T, we can specify a 'null value' T_0 such that values of T close to T_0 present the strongest evidence in favor of the null hypothesis, whereas values of T far from T_0 present the strongest evidence against the null hypothesis.

Total sum of squares	In statistical data analysis the total sum of squares is a quantity that appears as part of a standard way of presenting results of such analyses. It is defined as being the sum, over all observations, of the squared differences of each observation from the overall mean.
	In statistical linear models, (particularly in standard regression models), the TSS is the sum of the squares of the difference of the dependent variable and its grand mean: $\sum_{i=1}^{n} (y_i - \bar{y})^2$.
	For wide classes of linear models: Total sum of squares = explained sum of squares + residual sum of squares.
T-test	A t-test is any statistical hypothesis test in which the test statistic follows a Student's t distribution if the null hypothesis is true. It is most commonly applied when the test statistic would follow a normal distribution if the value of a scaling term in the test statistic were known. When the scaling term is unknown and is replaced by an estimate based on the data, the test statistic (under certain conditions) follows a Student's t distribution.
Confidence interval	In statistics, a confidence interval is a kind of interval estimate of a population parameter and is used to indicate the reliability of an estimate. It is an observed interval (i.e. it is calculated from the observations), in principle different from sample to sample, that frequently includes the parameter of interest, if the experiment is repeated. How frequently the observed interval contains the parameter is determined by the confidence level or confidence coefficient.
Holm-Bonferroni method	In statistics, the Holm-Bonferroni method performs more than one hypothesis test simultaneously.
	Suppose there are k null hypotheses to be tested and the overall type 1 error rate is α. Start by ordering the p-values and comparing the smallest p-value to α/k. If that p-value is less than α/k, then reject that hypothesis and start all over with the same α and test the remaining k − 1 hypothesis, i.e. order the k − 1 remaining p-values and compare the smallest one to α/(k − 1). Continue doing this until the hypothesis with the smallest p-value cannot be rejected. At that point, stop and accept all hypotheses that have not been rejected at previous steps.
Carlo Emilio Bonferroni	Carlo Emilio Bonferroni was an Italian mathematician who worked on probability theory. Carlo Emilio Bonferroni was born in Bergamo on 28 January 1892 and died on 18 August 1960 in Firenze (Florence). He studied in Torino (Turin), held a post as assistant professor at the Turin Polytechnic, and in 1923 took up the chair of financial mathematics at the Economics Institute in Bari.

Chapter 14. Comparing Groups: Analysis of Variance Methods

Exploratory data analysis	In statistics, exploratory data analysis is an approach to analyzing data sets to summarize their main characteristics in easy-to-understand form, often with visual graphs, without using a statistical model or having formulated a hypothesis. Exploratory data analysis was promoted by John Tukey to encourage statisticians visually to examine their data sets, to formulate hypotheses that could be tested on new data-sets. Tukey's championing of EDA encouraged the development of statistical computing packages, especially S at Bell Labs: The S programming language inspired the systems 'S'-PLUS and R. This family of statistical-computing environments featured vastly improved dynamic visualization capabilities, which allowed statisticians to identify outliers and patterns in data that merited further study.
Multiple comparisons	In statistics, the multiple comparisons or multiple testing problem occurs when one considers a set of statistical inferences simultaneously. Errors in inference, including confidence intervals that fail to include their corresponding population parameters or hypothesis tests that incorrectly reject the null hypothesis are more likely to occur when one considers the set as a whole. Several statistical techniques have been developed to prevent this from happening, allowing significance levels for single and multiple comparisons to be directly compared.
Tukey method	The Tukey method, is a single-step multiple comparison procedure which applies simultaneously to the set of all pairwise comparisons $\mu_i - \mu_j$. The confidence coefficient for the set, when all sample sizes are equal, is exactly $1 - \alpha$. For unequal sample sizes, the confidence coefficient is greater than $1 - \alpha$. In other words, the Tukey method is conservative when there are unequal sample sizes.
Data analysis	Analysis of data is a process of inspecting, cleaning, transforming, and modeling data with the goal of highlighting useful information, suggesting conclusions, and supporting decision making. Data analysis has multiple facets and approaches, encompassing diverse techniques under a variety of names, in different business, science, and social science domains. Data mining is a particular data analysis technique that focuses on modeling and knowledge discovery for predictive rather than purely descriptive purposes.
Regression analysis	In statistics, regression analysis includes many techniques for modeling and analyzing several variables, when the focus is on the relationship between a dependent variable and one or more independent variables. More specifically, regression analysis helps one understand how the typical value of the dependent variable changes when any one of the independent variables is varied, while the other independent variables are held fixed.

Ronald Aylmer Fisher	Sir Ronald Aylmer Fisher FRS (17 February 1890 - 29 July 1962) was an English statistician, evolutionary biologist, eugenicist and geneticist. He was described by Anders Hald as 'a genius who almost single-handedly created the foundations for modern statistical science,' and Richard Dawkins described him as 'the greatest of Darwin's successors'.
Mains	Mains is the general-purpose alternating current (AC) electric power supply. The term is not often used in the United States and Canada. In the US, Mains power is referred to by a variety of formal and informal names, including household power, household electricity, domestic power, wall power, line power, AC power, city power, and grid power.
Main effect	In the design of experiments and analysis of variance, a main effect is the effect of an independent variable on a dependent variable averaging across the levels of any other independent variables. The term is frequently used in the context of factorial designs and regression models to distinguish main effects from interaction effects. For example, in factorial designs, the main effect is what the independent variables elicit when averaged out over each other.
Inference	Inference is the act of drawing a conclusion by deductive reasoning from given facts. The conclusion drawn is also called an inference. The laws of valid inference are studied in the field of logic.
Factorial	In mathematics, the factorial of a positive integer n, denoted by n!, is the product of all positive integers less than or equal to n. For example, $5! = 5 \times 4 \times 3 \times 2 \times 1 = 120$ 0! is a special case that is explicitly defined to be 1. The factorial operation is encountered in many different areas of mathematics, notably in combinatorics, algebra and mathematical analysis.
Ranking	A ranking is a relationship between a set of items such that, for any two items, the first is either 'ranked higher than', 'ranked lower than' or 'ranked equal to' the second. In mathematics, this is known as a weak order or total preorder of objects. It is not necessarily a total order of objects because two different objects can have the same ranking.
Financial statement	A Financial statement is a formal record of the financial activities of a business, person, or other entity. In British English--including United Kingdom company law--a Financial statement is often referred to as an account, although the term Financial statement is also used, particularly by accountants.

For a business enterprise, all the relevant financial information, presented in a structured manner and in a form easy to understand, are called the Financial statements. They typically include four basic Financial statements:

· Balance sheet: also referred to as statement of financial position or condition, reports on a company's assets, liabilities, and Ownership equity at a given point in time. · Income statement: also referred to as Profit and Loss statement , reports on a company's income, expenses, and profits over a period of time.

| P-value | In statistical significance testing, the p-value is the probability of obtaining a test statistic at least as extreme as the one that was actually observed, assuming that the null hypothesis is true. In this context, value a is considered more 'extreme' than b if a is less likely to occur under the null. One often 'rejects the null hypothesis' when the p-value is less than the significance level α, which is often 0.05 or 0.01. When the null hypothesis is rejected, the result is said to be statistically significant. |

| Reaction time | Reaction time is the elapsed time between the presentation of a sensory stimulus and the subsequent behavioral response. It is often used in experimental psychology to measure the duration of mental operations, known as mental chronometry. The behavioral response is often a button press but can also be an eye movement, a vocal response, or some other observable behavior. |

| Estimation | In project management (i.e., for engineering), accurate estimates are the basis of sound project planning. Many processes have been developed to aid engineers in making accurate estimates, such as•Analogy based estimation•Compartmentalization (i.e., breakdown of tasks)•Delphi method•Documenting estimation results•Educated assumptions•Estimating each task•Examining historical data•Identifying dependencies•Parametric estimating•Risk assessment•Structured planning

Popular estimation processes for software projects include:•Cocomo•Cosysmo•Event chain methodology•Function points•Program Evaluation and Review Technique (PERT)•Proxy Based Estimation (PROBE) (from the Personal Software Process)•The Planning Game (from Extreme Programming)•Weighted Micro Function Points (WMFP)•Wideband Delphi. |

| Median | In probability theory and statistics, a median is described as the numerical value separating the higher half of a sample, a population, or a probability distribution, from the lower half. The median of a finite list of numbers can be found by arranging all the observations from lowest value to highest value and picking the middle one. |

Interquartile range	In descriptive statistics, the interquartile range also called the midspread or middle fifty, is a measure of statistical dispersion, being equal to the difference between the upper and lower quartiles. $IQR = Q_3 - Q_1$
	Unlike (total) range, the interquartile range is a robust statistic, having a breakdown point of 25%, and is thus often preferred to the total range.
	The IQR is used to build box plots, simple graphical representations of a probability distribution.
Sign test	In statistics, the sign test can be used to test the hypothesis that there is 'no difference in medians' between the continuous distributions of two random variables X and Y, in the situation when we can draw paired samples from X and Y. It is a non-parametric test which makes very few assumptions about the nature of the distributions under test - this means that it has very general applicability but may lack the statistical power of other tests such as the paired-samples t-test or the Wilcoxon signed-rank test.
	Let $p = Pr(X > Y)$, and then test the null hypothesis $H_0: p = 0.50$. In other words, the null hypothesis states that given a random pair of measurements (x_i, y_i), then x_i and y_i are equally likely to be larger than the other.
	To test the null hypothesis, independent pairs of sample data are collected from the populations $\{(x_1, y_1), (x_2, y_2), .$
Internet	The Internet is a global system of interconnected computer networks that use the standard Internet Protocol Suite (TCP/IP) to serve billions of users worldwide. It is a network of networks that consists of millions of private, public, academic, business, and government networks, of local to global scope, that are linked by a broad array of electronic and optical networking technologies. The Internet carries a vast range of information resources and services, such as the inter-linked hypertext documents of the World Wide Web (WWW) and the infrastructure to support electronic mail.
Browsing	Browsing is a kind of orienting strategy. It is supposed to identify something of relevance for the browsing organism. When used about human beings it is a metaphor taken from the animal kingdom.
Binomial	In elementary algebra, a Binomial is a polynomial with two terms--the sum of two monomials--often bound by parenthesis or brackets when operated upon. It is the simplest kind of polynomial other than monomials.
	· The Binomial $a^2 - b^2$ can be factored as the product of two other Binomials:

$a^2 - b^2 = (a + b)(a - b)$.

This is a special case of the more general formula:

$$a^{n+1} - b^{n+1} = (a - b) \sum_{k=0}^{n} a^k b^{n-k}$$

· The product of a pair of linear Binomials (ax + b) and (cx + d) is:

(ax + b)(cx + d) = acx^2 + axd + bcx + bd.

· A Binomial raised to the nth power, represented as

$(a + b)^n$

can be expanded by means of the Binomial theorem or, equivalently, using Pascal's triangle. Taking a simple example, the perfect square Binomial $(p + q)^2$ can be found by squaring the first term, adding twice the product of the first and second terms and finally adding the square of the second term, to give $p^2 + 2pq + q^2$.

· A simple but interesting application of the cited Binomial formula is the '(m,n)-formula' for generating Pythagorean triples: for m < n, let a = $n^2 - m^2$, b = 2mn, c = $n^2 + m^2$, then $a^2 + b^2 = c^2$.

Binomial distribution	In probability theory and statistics, the binomial distribution is the discrete probability distribution of the number of successes in a sequence of n independent yes/no experiments, each of which yields success with probability p. Such a success/failure experiment is also called a Bernoulli experiment or Bernoulli trial; when n = 1, the binomial distribution is a Bernoulli distribution. The binomial distribution is the basis for the popular binomial test of statistical significance.
Wilcoxon signed-ranks test	Wilcoxon signed-ranks test a non-parametric alternative to the paired Student's t-test for the case of two related samples or repeated measurements on a single sample. It involves comparisons of differences between measurements, so it requires that the data are measured at an interval level of measurement.

Chi-squared test	A chi-squared test, also referred to as chi-square test or χ^2 test, is any statistical hypothesis test in which the sampling distribution of the test statistic is a chi-squared distribution when the null hypothesis is true, or any in which this is asymptotically true, meaning that the sampling distribution (if the null hypothesis is true) can be made to approximate a chi-squared distribution as closely as desired by making the sample size large enough. Some examples of chi-squared tests where the chi-squared distribution is only approximately valid:•Pearson's chi-squared test, also known as the chi-squared goodness-of-fit test or chi-squared test for independence. When mentioned without any modifiers or without other precluding context, this test is usually understood .•Yates's correction for continuity, also known as Yates' chi-squared test.•Cochran-Mantel-Haenszel chi-squared test.•McNemar's test, used in certain 2 × 2 tables with pairing•Linear-by-linear association chi-squared test•The portmanteau test in time-series analysis, testing for the presence of autocorrelation•Likelihood-ratio tests in general statistical modelling, for testing whether there is evidence of the need to move from a simple model to a more complicated one (where the simple model is nested within the complicated one). One case where the distribution of the test statistic is an exact chi-squared distribution is the test that the variance of a normally distributed population has a given value based on a sample variance.
Null hypothesis	The practice of science involves formulating and testing hypotheses, assertions that are falsifiable using a test of observed data. The null hypothesis typically corresponds to a general or default position. For example, there is no relationship between two measured phenomena, or a potential treatment has no effect.
Correlation	In statistics, correlation (often measured as a correlation coefficient, ρ) indicates the strength and direction of a relationship between two random variables. The commonest use refers to a linear relationship. In general statistical usage, correlation or co-relation refers to the departure of two random variables from independence.
Independence	In probability theory, to say that two events are independent intuitively means that the occurrence of one event makes it neither more nor less probable that the other occurs. For example:•The event of getting a 6 the first time a die is rolled and the event of getting a 6 the second time are independent.•By contrast, the event of getting a 6 the first time a die is rolled and the event that the sum of the numbers seen on the first and second trials is 8 are not independent.•If two cards are drawn with replacement from a deck of cards, the event of drawing a red card on the first trial and that of drawing a red card on the second trial are independent.•By contrast, if two cards are drawn without replacement from a deck of cards, the event of drawing a red card on the first trial and that of drawing a red card on the second trial are again not independent.

Similarly, two random variables are independent if the conditional probability distribution of either given the observed value of the other is the same as if the other's value had not been observed. The concept of independence extends to dealing with collections of more than two events or random variables.

Multiple correlation

In statistics, multiple correlation is a linear relationship among more than two variables. It is measured by the coefficient of multiple determination, denoted as R^2, which is a measure of the fit of a linear regression. A regression's R^2 falls somewhere between zero and one (assuming a constant term has been included in the regression); a higher value indicates a stronger relationship among the variables, with a value of one indicating that all data points fall exactly on a line in multidimensional space and a value of zero indicating no relationship at all between the independent variables collectively and the dependent variable.

Standardized

In mathematical statistics, a random variable X is standardized using the theoretical (population) mean and standard deviation:

$$Z = \frac{X - \mu}{\sigma}$$

where $\mu = E(X)$ is the mean and σ = the standard deviation of the probability distribution of X.

If the random variable under consideration is the sample mean:

$$\bar{X} = \frac{1}{n} \sum_{i=1}^{n} X_i$$

then the standardized version is

$$Z = \frac{\bar{X} - \mu}{\sigma/\sqrt{n}}.$$

Logistic regression

In statistics, logistic regression is a type of regression analysis used for predicting the outcome of a categorical (a variable that can take on a limited number of categories) criterion variable based on one or more predictor variables. Logistic regression can be bi- or multinomial. Binomial or binary logistic regression refers to the instance in which the criterion can take on only two possible outcomes (e.g., 'dead' vs. 'alive', 'success' vs. 'failure', or 'yes' vs. 'no').

| Sampling distribution | In statistics, a sampling distribution is the probability distribution of a given statistic based on a random sample. Sampling distributions are important in statistics because they provide a major simplification on the route to statistical inference. More specifically, they allow analytical considerations to be based on the sampling distribution of a statistic, rather than on the joint probability distribution of all the individual sample values. |

1. In mathematics, the _____ of a positive integer n, denoted by n!, is the product of all positive integers less than or equal to n. For example, $5! = 5 \times 4 \times 3 \times 2 \times 1 = 120$

   0! is a special case that is explicitly defined to be 1.

   The _____ operation is encountered in many different areas of mathematics, notably in combinatorics, algebra and mathematical analysis.

   a. Factorial moment
   b. Factorial prime
   c. Factorial
   d. Genocchi number

2. _____ is the act of drawing a conclusion by deductive reasoning from given facts. The conclusion drawn is also called an _____. The laws of valid _____ are studied in the field of logic.

   a. Abduction
   b. Analytic reasoning
   c. Automated reasoning
   d. Inference

3. In statistics, the _____ or multiple testing problem occurs when one considers a set of statistical inferences simultaneously. Errors in inference, including confidence intervals that fail to include their corresponding population parameters or hypothesis tests that incorrectly reject the null hypothesis are more likely to occur when one considers the set as a whole. Several statistical techniques have been developed to prevent this from happening, allowing significance levels for single and _____ to be directly compared.

   a. closed linear operators
   b. George Berkeley
   c. Cristina Bicchieri
   d. Multiple comparisons

4. In probability theory and statistics, the _____ is a measure of how far a set of numbers is spread out. It is one of several descriptors of a probability distribution, describing how far the numbers lie from the mean (expected value). In particular, the _____ is one of the moments of a distribution.

    a. Variation ratio
    b. Variogram
    c. Variance
    d. Full width at half maximum

5. In statistics, _____ (often measured as a _____ coefficient, ρ) indicates the strength and direction of a relationship between two random variables. The commonest use refers to a linear relationship. In general statistical usage, _____ or co-relation refers to the departure of two random variables from independence.

    a. Correlation
    b. Pearson product-moment correlation coefficient
    c. Sample covariance
    d. Sample covariance matrix

1. c
2. d
3. d
4. c
5. a

You can take the complete Chapter Practice Test

for Chapter 14. Comparing Groups: Analysis of Variance Methods
on all key terms, persons, places, and concepts.

Online 99 Cents

http://www.epub27.14.20551.14.cram101.com/

Use www.Cram101.com for all your study needs

including Cram101's online interactive problem solving labs in

chemistry, statistics, mathematics, and more.

Other Cram101 e-Books and Tests

Want More?
Cram101.com...

Cram101.com provides the outlines and highlights of your
textbooks, just like this e-StudyGuide, but also gives you the
PRACTICE TESTS, and other exclusive study tools for all of your
textbooks.

Learn More. *Just click*
http://www.cram101.com/

CPSIA information can be obtained at www.ICGtesting.com
Printed in the USA
LVOW02s1807080813

346998LV00004B/680/P